Acceptance and Commitment & Dialectical Behavior Therapy Made Simple

Practical ACT & DBT Guide for Learning Mindfulness and Emotion Regulation

Leona S. Murray

Summary

Acceptance and Commitment Therapy

Dialectical Behavior Therapy

Page intentionally left blank

Acceptance and Commitment Therapy

The Action-Oriented Psychotherapeutic Approach of ACT to Reducing Stress, Anxiety, Anger, Panic Attacks, & Depression

Leona S. Murray

This edition is totally revised and edited, to guarantee contents' quality.

Did you know that can download the audiobook of this book for free?
Click here for Audible US
Click here for Audible UK
Click here for Audible FR
Click here for Audible DE

Don't forget to leave a review on this book. Is simple! Just click on this LINK and you will be directed to the right page. It's very important for me. I will appreciate if you do.

Introduction

A therapist realizes that everyone is unique and that the level of their challenges in life fluctuates. This further implies that there is no one-size-fits-all kind of arrangement; however, having the option to assist another person in overcoming such troubles with the end goal for them to meet their maximum capacity in life is undoubtedly an extraordinary method for creating an enormous effect on humanity. If you're willing to help somebody to roll out a significant improvement in their life, whether you are a social specialist, a school advocate, or maybe even a clinician needing to add to your accreditations, you may consider knowing about the acceptance and commitment therapy, otherwise called ACT.

ACT is viewed as the third wave behavioral therapy, after mindfulness-based cognitive therapy and rationalistic conduct therapy. Empowering an individual to perceive their values, to follow up on them, and, in doing so, contemporary significance and vitality in life while expanding one's psychological adaptability is the primary reason for this therapy. Instead of cognitive behavioral therapy, ACT doesn't show individuals how they can control their feelings, emotions, or even to avoid any unwanted thoughts. Rather, it enables them to acknowledge and build up a new careful relationship with their encounters so that they can take action that is in line with their values.

As you go further into ACT, you will find that it chips away at six center procedures. They are:

1. Cognitive defusion - acknowledging feelings and thoughts for what they indeed are
2. Values - finding what is generally critical to the person
3. Acceptance - rehearsing non-critical attention to encounter and also to push ahead
4. Self-as-setting - getting careful and being secure in the extraordinary feeling of self
5. Contact with right now - getting mindful of the present time and place with transparency
6. Committed action - thinking of objectives dependent on one's values and actualizing them in a capable way

ACT is an observationally upheld therapy. Among the issues wherein ACT has been actualized as treatment include smoking, negative self-perception, psychosis, dietary problems, and work environment stress. Aside from that, it was likewise utilized as a sort of training model. Without a doubt taking up Acceptance and Commitment Therapy preparing will make you equipped to help people who manage different challenges and make improvements in their satisfaction.

The Process of Acceptance and Commitment Therapy

Robust psychological appraisal which covers a definite practical investigation of the introducing issue. Think of typical examples of symptoms, thoughts, feelings, and social responses. They are mainly searching for avoidant reacting and inefficient endeavors in controlling the experience that brings about not seeking after esteemed objectives. Additionally, evaluation covers social conditions, substantial impacts on current practices. The demonstrative appraisal, like other psychological treatments, is viewed as less significant. The key emphasis is on the 'usefulness' of the techniques which an individual uses to carry on with their life/take care of their issues and the practical examination of current conduct, thoughts, and feelings.

Case Conceptualization

In contrast to Cognitive Therapy, ACT advisors, once in a while, present a specific case plan once in a while. Instead, the advisor utilizes allegories to depict the ordinary useful qualities that the customer offers and inquires as to whether this thought fits with their direct experience. Models incorporate the cruising pontoon allegory, the travelers on the transport illustration, the man in the opening representation, and the back-and-forth with a beast similitude.

Imaginative Hopelessness

Right now, we are working with the customer to help them analyze the 'usefulness' of the procedures they use. The main objective right now is to look at what the customer has been doing, how well it has worked, and to check whether these procedures may be a part of the issue. Focal messages that endeavors to control encounters might be a piece of the problem instead of the arrangement. Further significant words show that the customer isn't broken, merely using unworkable procedures to live with what life gives them (or has given them). The design is to perceive that business, as usual, probably won't be useful, that this therapy should be something new and distinctive to what they've previously done, and that relinquishing unworkable systems may mean developing an eagerness to sit with what is troublesome or difficult; altogether, new methodologies can be attempted to 'live with' the issue. Although these three starting stages are presented in this synopsis as isolated stages, they can be mixed, contingent upon the customer and specialist.

Accomplishing the ACT work

When the customer and advisor have formed an effective working partnership, and there is understanding that what has gone before has not functioned admirably and that the customer expresses readiness to try to let go of vain techniques. The principle work of ACT becomes about characterizing what the customers need to be about (their qualities), making stable arrangements, and conducting steps in those ways. As this occurs, the customer's psyche will begin to think of reasons why this can't happen, wishes to maintain a strategic distance from it, and in combination with thoughts, convictions, and rules about how life ought to be lived, about the idea of their concern about themselves as well as others. These mostly serve as deterrents between the individual and their qualities.

We use analogy, sympathy, demonstrating of acceptance, experiential activities, and care procedures to try to cultivate eagerness to have these deterrents come up, similarly as they seem to be. The point is to develop a perceptive and isolate an inviting position towards these hindrances. We use social mediations, for example, progressions and objective setting, tying, demonstrating, pretending, and practicing to attempt to display, start, and strengthen practices that move the individual towards their esteemed objectives. We use the remedial space to allow individuals to segregate practices and conditions that lead them away from their qualities and try to give individuals a more prominent ability to choose in those minutes.

We have to begin with values for satisfied customers: what are we going to do for this odd thing called therapy? It has to be something that they find problematic. Numerous individuals are far from the way they'd like their lives to be, and this work can be excruciating and slow. This includes a lot of acceptance and defusion regularly just to connect with what is essential and permit the likelihood that the therapy could be attached to working for it. Different customers are so melded with thoughts and convictions about themselves that they have to begin with defusion, acceptance, and genuinely reaching the unworkability of their techniques and the unhelpfulness of 'over-thinking' and of purchasing thoughts.

Defusion Based Interventions
Unlike cognitive therapy, thoughts are not classified as a basis, silly, mutilated, useful, or useless. Instead, we see the relationship one has with those thoughts as either functional or unworkable at whatever minute. We don't truly get into examining how much a customer accepts or doesn't accept an idea or conviction, and we don't get involved in discussing whether the concept is valid or exact. Instead, we use care and symbolism based activities, just like other experiential exercises, to assist individuals with considering thoughts to be thoughts, including those that are incredibly agonizing.

We do this in the administration of unfastening thoughts as explanations behind the activity, all together that activities become all the more reliably controlled by the person's decisions towards their qualities, as opposed to what their 'minds' disclose to them that they should do. We allude to minds as psyches to recognize the intention of self. We call attention to the development of language, for example, the historical backdrop of a word, with an end goal to consider words to be words, in any event, when they are about troublesome things. We use physicalizing motions and pictures to describe thoughts and convictions as a significant aspect of oneself, yet separate from oneself.

Qualities Based Interventions

While numerous individuals may set objectives for themselves in specific issues daily, a significant number of us live with an understood feeling of what is critical to us. Different methods for saying this are, "What would we like to be? What matters most to us?" For some individuals, life is so distant from these thoughts that they are exceptionally outsiders and very compromising. For some of us, life is tied in with doing what we are told, what is expected from us, and what fits in with our jobs. This can be to such an extent that in various parts of our lives, we are not, in any case, mindful of picking, or aware that a decision even exists. Indeed, even individuals who are working effectively on the planet might not have thought about this part of their lives (including the individuals we call advisors).

Right now, we 're working to help individuals reach those inquiries: What would I like to stand for? On the off chance that I found a good pace, what would my life be about? What is the main thing for me? We do this through symbolism exercises, sitting in the inquiry, and investigating it. By the complicated procedure of seeing what is excruciating or that we maintained a strategic distance from. Qualities and vulnerabilities are poured from a similar vessel: in our agony, we find what is valuable. This is frequently about the association with others, feeling we have a place and are cherished, that we have any effect on the individuals around us, that we matter, that we care for individuals in our lives, that we contribute here and there to our general surroundings.

We attempt to enable the individual to reach this in their life, and the therapy turns out to be entirely in the administration of carrying on with an esteemed life, with less battling. Usually, this will include working skilfully with the impediments typically emerging as the individual makes those moves, relinquishing techniques that don't work, and helping customers to be happy with whatever comes up as they move towards values. Some portion of the work could be linked in with concocting actual conduct objectives by these qualities and breaking these into little strides to permit a progression of littler moves to major moves. The entirety of the thoughts, recollections convictions, and everything the customer has been battling with will need to join the party.

The ACT Therapeutic Stance
In this synopsis it may be evident that the remedial position in the ACT is one of correspondence. The specialist and customer are both minded individuals. They are the two individuals who have now and again known battle, grief, happiness, trouble, aching, and have harmed. They are the two individuals who have acted in a manner that they lament or were not their best selves, and the two got captured into rules of how we 'should' be living and what we 'should' do at whatever minute. The ACT specialist invites this ordinary humanity and makes endeavors in the relationship to build up a correspondence. This may include chosen self-revelation, in administering the customer's qualities or battles. A capable ACT specialist is likely to make a careful, defused, adaptable space, in which anything can be examined in the room, particularly

processing issues concerning what is being experienced by the two gatherings, in the present time and place.

Simultaneously, the ACT advisor is generally educational or controlling, in that they are there for the administration of the customer's increasingly active esteemed living. Customers may require a great deal of consolation and restorative impact to surrender old examples, attempt new things, and sit with troublesome material. The ACT specialist is dynamic, and objective situated, however, if the goal is to be better, ready to sit eagerly with disturbing content, the advisor may sit discreetly, understandingly, tenderly, permitting space for feeling to be unequivocally present and helping the customer in developing a tolerating, adaptable, defused and present-focused position towards that problematic material. Off chance that once again, the immediate therapeutic objective is to hold a troublesome idea all the more gently, the specialist might be lively, utilizing humor, utilizing games and activities that permit the customer to encounter the concept as an idea and might be very disrespectful towards the idea. This will consistently rely on the customer and the idea of the relationship and is never done to put down the customer.

Know More About Psychotherapy

You may have heard the term 'psychotherapy' and have pondered precisely what it implies. Numerous therapists, clinicians, and emotional wellness experts use terminologies such as psychotherapy and cognitive conduct; anyway, the healthy individual is curious about the significance of these brain science-related words. Psychotherapy and therapy are occasionally utilized conversely. The objectives of psychotherapy are to:

1. increase a person's feeling prosperity
2. improve emotional wellness
3. improve connections

There is a wide range of strategies and techniques that psychotherapists use to accomplish these objectives. A few procedures include:

1. relationship building
2. discourse
3. correspondence
4. conduct change
5. pretense

Similarly, as with any expert in the field of brain science, all conversations you have with a psychotherapist are secret. Privacy is important to brain science as it is identified with trust, receptiveness, and polished methodology.

There are various ways of thinking in regards to psychotherapy and relying upon your therapists' instructive foundation and convictions, and they may use at least one methodology:

Psychoanalytic - because of the recovery of thoughts, including free affiliations, and dreams.

1. Conduct Therapy - centers around changing examples of conduct to improve enthusiastic responses and interactions with others

2. Cognitive Behavioral - centers around recognizing and impacting or changing damaging or useless convictions, reactions, feelings, and practices

3. Psychodynamic - centers around uncovering the oblivious substance of the mind (intuitive) with an end goal to ease psychological pressures

4. Existential - a philosophical methodology dependent on the existential conviction that people are separated from everyone else on the planet, which prompts feelings of

pointlessness which are overwhelmed by creating or recreating one's values and implications

5. The humanistic - an existential methodology that accepts that human understanding depends on the improvement of the person with an accentuation on emotional importance and worry for constructive development.

Psychotherapy can help recognize and treat various emotional wellness issues, including (yet not constrained to):

1. Depression
2. Anxiety
3. Eating disorders
4. Addiction and alcoholism
5. Bipolar disorder
6. A lack of ability to concentrate consistently Disorder and Attention Deficit Hyperactive Disorder
7. Over the top Compulsive Disorder
8. Post Traumatic Stress Disorder

You should contact your local emotional wellbeing specialist on the off chance you're thinking about psychotherapy or might want to study therapy and how it can support you or your family. Numerous psychotherapists acknowledge protection plans or offer variable expense rates with the goal of moderate help. It takes mental fortitude to look for the assistance of an expert therapist. However, you will find that once you do, the procedure can be agreeable and fulfilling.

Have You Discovered What You Value Most in Your Life?

Our values and convictions change depending on circumstances and point of view. What makes the most difference to you might be completely different to what makes the most difference to another; while you may value your profession more than your partner because your vacation satisfies your needs; to endure, be independent and financially successful, your partner may value your relationship more because the parts of adoration found in the relationship, for example, friendship, love, to cherish and be adored, are characteristics that stand out. While you may give all your opportunity to making progress toward being the best individual, your companion may decide to give their opportunity to building kinships because the feeling of achievement that matters most to you isn't as important to them as friendship and association with individuals. Neither of the values is correct nor wrong, and they are basically in an alternate request of need. Therefore, the most important thing is to comprehend what is the most important thing to you since this what brings individual and life satisfaction. Find what makes the most difference to you now.

Make a list of the things that currently matter to you. (For instance, dependability, friendship, love, internal harmony, endurance, great wellbeing and wellness, balance, delight, satisfaction, money-related freedom, achievement, progression, otherworldly advancement, graciousness, helping others, acceptance, understanding or self-awareness, and improvement).

Think about why these values matter to you - This causes you to characterize the distinction between what you think matters and the main thing.

Put your values in the present request of need.

Investigate your life now and check if your way of living and being aligns with your values.

Take your most important value and make a rundown of things you can do to bring a greater amount of this quality into your day-to-day life. On the off chance that you have picked euphoria, at that point, what would you be able to do right presently to create more satisfaction? On the off chance that you have picked achievement, what would you be able to do right currently to create further achievement? On the off chance that you have picked friendship, at that point, what would you be able to do to help create current relationships or start new ones? Whatever value you intend to create, make a rundown of twelve alternatives that can help you accomplish it, and don't surrender the conceptualizing until you have completed this undertaking. When you are clear about what makes a difference the most, you have the establishments for creating a method for living and being that holds importance and value to you; this leads to the establishment of individual satisfaction.

By perceiving our values and being consistent with them, and perceiving that others also value theirs in an alternate request, we come to understand, endure, and acknowledge one another. At the point when we are content, which is a consequence of living in arrangement with our values, we have no compelling reason to pass judgment on others, and without equity, there is harmony. Right now, it could state that the guilty pleasure of self-revelation adds to world harmony.

Chapter 1

For many years, researchers in the field of psychology have worked to promote science-based, time-limited human health interventions that seek to overcome intellectual health conditions. As a result, many people have had full-size success in addressing and managing various issues, and as a result, the process is going well. Nevertheless, long-term recovery and relapse prevention remain considerable as areas of notable concern for those seeking remedies for intellectual health conditions. More recently, new types of therapies, such as the ACT, have been developed to increase long-term success in therapy in the event of intellectual conditions.

The Act is based on Relational Frame Theory (RFT), a research faculty focusing on human language and cognition. RFT suggests that rational competencies using human thought to overcome troubles may be ineffective in overcoming psychological pain. Based on this suggestion, the ACT measure was developed to educate people that even though psychological pain is common, we can research ways to stay healthy, how we think of pain, and live life by completing it.

Starting in the late 1990s, more than one comprehensive measure manual has been developed to outline how the ACT is used to treat various intellectual health conditions. The use of these manuals has been empirically researched and has produced a guide to the use of ACT in the treatment of substance abuse, psychosis, anxiety, depression, chronic pain, and disorders.

Principle of act

ACT theory no longer outlines unwanted emotional experiences as signs of problems. It works as an alternative to seeing some people's tendency to find therapy as damaged or unsuitable and aiming to help people recognize the fullness and vitality of life. This fullness encompasses a broad spectrum of human experience, which inevitably includes pain with certain conditions.

Accepting things as they come, in addition to evaluating them or trying to change them, is a skill developed through the Mindfulness Workout routine within and outside the session. The ACT no longer attempts to immediately exchange or stop unwanted thoughts or feelings (as does cognitive behavior therapy), although alternatively encouraging a human to foster a new and compassionate relationship with these experiences. This shift can relieve humans from the difficulties of manipulating their experiences and also help them to be extra open to moving regularly with their values, and explanations of values and definition of desires based on values.

Six core procedures of the act

Psychological flexibility, the major objective of ACT, usually comes about by several main processes.

Developing creative despair requires using previous efforts to heal or overcome these difficulties in bringing a character to therapy. The ACT creates the possibility for people to work more regularly with what is most important to them by paying attention to the lack of functionality or practicality of these efforts.

Accepting a person's emotional journey can be described as a system of mastery for experiencing human emotions differently with a one-way, open, and accepting attitude.

Choosing valuable life directions is the process of clarifying what is most important in life and how one wants to live life.

Taking action can refer to the commitment to make an adjustment and interaction during the most valuable behavior.

These approaches are overlapping and interconnected, not separate. All of these methods are brought to and developed through direct experiences that are recognized and taken through characterization in therapy during treatment. Psychological resilience can only be defined as "the ability to be present, open, and do".

MINDFULNESS and ACT

Mindfulness can be described as maintaining contact with the current moment as compared to flowing in an automatic pilot. Mindfulness allows a person to associate with seeing oneself as a distinct segment of the conscious self. Mindfulness methods regularly help humans raise awareness of the five senses as well as their thoughts and feelings.

Mindfulness will also enhance a person's ability to separate from thoughts. Challenges associated with painful feelings or situations are often reduced first and then eventually accepted. Acceptance is the ability to enable internal and external experience or to overcome experience rather than hostility. If a person thinks, "I am a terrible person," the person will likely be requested to say, "I think that I am a terrible person." It separates the person from the sensation, from which he is separated. It has a negative charge.

When humans travel with painful feelings, such as anxiety, they may be advised to open up, breathe, or make room for anxiety's physical feeling and allow it to remain there, as if it were sharp. Or to reduce.

Uses Values and Acts

Value explanations can help humans tell what's most important — their values, in other words, take tremendous action directed by these values. A mental fitness professional will typically use a variety of workouts to help make them aware of the chosen values. These values often act as a compass in the path of deliberate and exceptional behavior.

Exploring painful thoughts or overcoming a problem may also interfere with the ability to execute purposeful, value-driven action. Through this challenge-free mind, ACT can help human beings act according to their values and live naturally and fully.

What is an ACT?

The ACT discipline no longer provides honorable certificates for physicians wishing to offer this type of therapy.

The Association for Contextual Behavior Science (ACBS) compiles a voluntary registry of members who have identified themselves as ACT physicians. This registry may also be an appropriate place to join in the search for a company ACT.

ACBS also offers the following suggestions for those wishing to find an ACT doctor:

Contact the Department of Psychology, Social Work, or Psychiatry at a nearby college or university. School members or personnel who are professionals in conducting treatment or cognitive-behavioral measures may also learn about a neighborhood ACT therapist.

The United States-based website of the Association for Behavioral and Cognitive Therapy compiles a list of behavioral and cognitive therapies providers. These physicians may also grant ACT, or they may even know a peer who supports ACT.

In international locations other than the United States, agencies similar to ABCT may also be the perfect location to search for vendors who deliver ACT or who can make referrals to ACT-trained therapists.

Acceptance and Commitment Therapy (ACT) gets its title from one of its core messages: Accepting something that is out of your non-public control and improving the pace that makes your life better and more prosperous.

ACT aims to maximize human viability for a prosperous, fulfilling, and vital life. The ACT (called as the phrase 'act,' now in initials) does not occur by:

a) Instructing your psychological abilities to deal with your painful thoughts and feelings effectively - in such a way that they have little impact on you (these are identified as mindfulness skills).

B) Helping you clarify what is important to you - that is, your values - then use that information to guide, encourage, and inspire you to alternate your life for the better.

Mindfulness is a "hot topic" in Western psychology right now - more and more known as a powerful therapeutic intervention for the whole, from work stress to frustration - and also as a useful tool for increasing emotional intelligence. Acceptance and commitment therapy is an effective mindfulness-based measure (and coaching model) that currently leads the discipline in terms of research, utility, and outcomes.

Mindfulness is an intellectual empire of awareness, meditation, and openness, which allows you to interact at any moment fully. In a state of mindfulness, hard thoughts and feelings have little effect on you and so on - so it is very beneficial for enhancing athletic or enterprise performance, ranging from a full-fledged psychopathic disease. In many fashions of coaching and therapy, mindfulness is specifically taught through meditation. However, in the ACT, meditation is considered to be only one of many ways to study these abilities - and this is a reasonable thing, due to the fact that most humans no longer like to meditate! ACT provides you with an important class of tools to study mindfulness capabilities - many of which require only a few minutes.

The ACT breaks down mindfulness capacity into three categories:

1) Misconception: Getting away from it, and letting go, beliefs, useless thoughts, and memories

2) Acceptance: Making room for painful urges, feelings, and sensations, and allowing them to come and go, leaving the conflict

3) Get in touch with the present moment: Completely enamored with your here and now experience, with an attitude of openness and curiosity

These three abilities require you to use an issue of your own for which no words exist in common everyday language. This is the part of you that is sufficient to generate awareness and attention. In the ACT, we often call it 'self-observation.' We can talk about 'self' in many ways, although in a day-to-day language, we mainly discuss your physical self, your body, and your self-thinking self. 'Observing oneself' is that phase of you that is in a position to take a look at both your physical self and your question self. A better word, in my view, is 'pure awareness' - because it is everything: just awareness, nothing else. It is the part of you that is aware of everything else: aware of every thought, each emotion, and everything you see, hear, touch, taste, smell, and do.

Acceptance and Commitment Therapy (ACT) is an innovative and unique approach to an alternative in practice that changes the very floor rules of most Western psychiatry. It is a mindfulness-based, value-oriented behavioral therapy, which has many similarities to Buddhism, yet is by no means non-secular; It is a contemporary scientific approach, based primarily on modern research in human behavioral psychology.

Period and distribution of the ACT

The ACT can be distributed in several types:

a) Long-term remedy - for example, spectrum protocol for ACT with borderline character disorder: 42-hour sessions.

B) Medium-term measures - for example, in JoAnne Dahl's protocol with persistent pain: eight hours total

C) Short therapy - for example, Patty Bak's protocol on the ACT with schizophrenia: only four 1-hour sessions

d) Ultra-brief measures - for example, in the most important care medical settings, Cancer Strosal work, place act can be of particularly high quality even in one or two to thirty-minute sessions.

ACT briefly

My style for both teaching and medicine is what I call the Simple Act Made Simple - a distilled, simplified model of ACT, which is strongly influenced by three factors:

1) The work of Kirk Strausshal, which is one of the promoters of the ACT, is noted for the simplicity, tenacity, and effectiveness of his interventions,

2) My previous training in solution-oriented counseling and various brief treatments, most of which can be tailored to this model, and

3) Most of all, according to my non-public motto: 'simplicity, clarity, accessibility.' The ACT was considering it in the

mid-eighties, so it comes with a splendid legacy of tools, interventions, techniques, and strategies. Instead, some of these are unnecessarily complex or long inverted.

I have used my personal mission to modify the tools, techniques and coaching methods, wherever possible, to make the act readily available to everyone - whether they are a health professional, physician, coach, students or not, customers, affected persons - or involved members of the popular public.

My personal desire in the ACT is to work quickly and effectively, resulting in the majority of my work falling below B), C) and D). However, I have some long-term customers with some extreme problems that I have seen for quite a few years. And this is one of the many cases that I love about the ACT model: its flexibility. I love how it can be easily customized and made as quick or as long as possible to suit the wishes of the customer, group, or organization.

Six key processes of psychological flexibility and act

The ACT has six main processes:

Psychologically projecting the potential of The Present Moment: consciously engaging with anything that is going on here.

Ability to step back or detach yourself from thoughts and worries and memories about disability: Instead of getting stuck in your thoughts, or pushed around through them, or struggling to get rid of them, you study how to let them come and go - as if they were just riding the last motor outside your house. You examine how to bring your thinking back, and you look at your thinking so that you can react successfully - rather than confusing or losing your thinking.

Skill to open and accept rooms for painful feelings and sensations. You do research to foment hostility with them, provide them with some breathing space, and let them get caught in it there, or be overwhelmed through them; The more you can open up, and provide room for them to move, the less difficult it is for your feelings to come and go, except to leave you or put you back.

The Observing Self is the part of you that is responsible for cognition and meditation. We do not have a word for it in the ordinary daily language - we usually speak of "mind." But there are two components of the mind: the astonished self - that is, the section that usually thinks; The part that accounts for all your thoughts, beliefs, memories, decisions, fantasies, etc. and then looking at yourself there - that phase of your thinking that is conscious of what you think or feel or do in a living situation or at any moment. Without it, you cannot increase these mindfulness skills. And the more you exercise these mindfulness skills, the more conscious you become of this stage of your mind, and it gets the right of entry when you want it. (The technical term for this, in ACT, is 'self-reference').

Values are what you favor about your lifestyle being deep in your heart. What do you want to stand for? What do you want to do with your time on this planet? Sooner or later, what matters to you in the big picture? What would you like to remember about the humans you love?

The committed pace of taking the pace of action is guided by the help of your values - what things to do - whether it is challenging or inconvenient.

You advance 'psychological' flexibility when putting all these matters together. It is the ability to be in the present moment with awareness and openness, and to take action, guided by using your values. In other words, it is the ability to exist, open up about what matters. The ability to increase your presence, and open up about what matters in your life - the greater your sense of vitality, well-being, and fulfillment.

More about the ACT model

ACT uses a variety of experimental workout routines to reduce the energy of adverse cognitive, emotional, and behavioral processes. It helps clients to essentially trade their relationship with painful thoughts and feelings, to pursue a transparent experience of themselves, to live in the present, and to take action, guided by their deepest values, in order to create a prosperous and important life.

The ACT holds that most psychological conflict occurs through the initial experiential avoidance, that is, by trying to avoid or gain unwanted personal experiences (such as unhelpful thoughts, feelings, sensations, urges, and memories). Clients' efforts in experiential avoidance may work over a quick period, although they regularly fail over the long term, and in the process, they often cause tremendous psychological pain. (For example, anticipating any serious addiction: in the short term, it gives you a proper feel and helps you get rid of unbearable thoughts and feelings - but in the long run, it destroys your health and vitality).

In the ACT, customers reinforce mindfulness capabilities (both traditional techniques, and many modern, revolutionary ones) that enable them to change their relationship with painful thoughts and feelings fundamentally. When customers use these competencies in daily life, painful feelings, and unexpected thoughts are a good deal and have an impact on them. Therefore, instead of wasting their time and power in a fruitless war with their inner experiences, they can invest their energy in taking the pace to direct their lifestyles higher using their deepest values.

The increasing rejuvenation of empirical data confirms that acceptance, mindfulness, and openness to ride are useful to treat depression, anxiety disorder, substance abuse, chronic pain, PTSD, anorexia, and even schizophrenia. The ACT is also a very useful mannequin for life coaching and executive coaching.

The ACT is one of the waves, 'third wave,' of behavioral therapy, as well as dialectical behavior therapy, mindfulness-based cognitive therapy, and many more. These treatment options include a movement in psychology that considers temperament and acceptance as necessary additions to change-oriented remedy strategies.

"What is mind?

We are halfway through a revolution. Science has been made great for centuries. External observation progresses our perception of the world. Physics revealed planetary action; chemistry is found to be an excellent factor case; biology has taught us how to understand and treat the disease. But during this time, there were many unanswered questions about something that might be even more important to us. That is the human mind. The thinking that is so difficult to study is contrary to the phenomenon as mentioned above, it is no longer something we can see, measure, or manipulate without problems. Moreover, recognition is the most complex entity in the universe.

Consider the following to give you a human experience of this complexity. The brain contains between ten billion and one hundred billion characters of nerve cells or neurons. Each of these neurons may have ten thousand connections to different neurons. This important mesh is the foundation of the mind and provides the upstairs jostle, all equally good and difficult to understand the perception of intellectual events, memory, and language.

The last several decades have seen the introduction of new applied sciences. Methods for reading this complicated organ. We feel more about it compared to the time which came before it in the last half-century. This period of rapid discovery coincides with making it larger in a wide variety. Different disciplines - many of them are brand new - find out about the mind. Since then,

a coordinated effort has been made between practitioners of these disciplines. This interdisciplinary approach has since been known as cognitive science. Unlike science before it, which once focused on the world outside, observable event or "outer space," this new activity sets the entire focus on exploring our intriguing intellectual world, or "inner space."

Cognitive science can be broadly expressed as an interdisciplinary scientific study of the mind. Its important methodology is the scientific method, although we will see that many other methods also contribute. A distinctive feature of cognitive science is its interdisciplinary approach. It results from the efforts of researchers working in a broad field. These include darshan. The field brings with it a unique set of tools and approaches that are significant goals of Psychology, Linguistics, Synthetic Intelligence, Robotics, and Neuroscience.

This ebook is meant to show that when it comes to analyzing something complex like the mind, no single approach is sufficient. Instead, we learn a lot from the collaboration between practitioners of these disciplines. The period for cognitive science is no longer too much for the sum of all these disciplines to complete work on their intersections or unique problems. In this understanding, cognitive science is no longer a unified discipline of study like every discipline itself, but a collaborative effort among researchers working in many fields. The glue that holds cognitive science collectively is a matter of mind; for the most part, the use of scientific methods. In conclusion, we also discuss how clearly cognitive science is integrated. Cognitive science is to identify what we really want.

Know what its theoretical view of thinking is. This approach is centered on computational thinking, which may alternatively be called information processing. Cognitive scientists view the mind as a record processor. Information processors have both attributes and changes in information. According to this view, there should be some form of mind, intellectual depiction, and techniques that act on and manipulate that information. These two ideas will be discussed in more detail later in this chapter.

Cognitive science is often credited with being influenced by the way it is pushed upward. The computer path consists of a data processor. Think about a personal computer for a minute. It performs several information processes. Information enters PCs by entering devices keyboard or modem. The facts can then be saved on the computer, for example, on a hard disk. Data may be processed by the use of software programs such as text editor. The results of this processing next may serve as a production, either for a reveal or a printer. Like fashion, we can assume that people are doing comparable work. Information is our "input" through mind perception - what we see or hear. It is stored in our memories processed into idea shapes.

Therefore, our thoughts can serve as the basis "Output," such as language or physical behavior. The direction of this analogy between human thought and computer systems is very much a high level of abstraction. The actual physical way in which information is stored in the computer bears much resemblance to human memory formation. But both systems are characterized by computations. In fact, it's no longer going some distance to say these cognitive scientists see the mind as a machine or system whose work they are trying to understand.
Representation

As mentioned earlier, illustration is fundamental to cognitive science. But what is a representation? Before listing the attributes of representation, it is helpful to describe four categories of representation briefly. A concept stands for an entity or group of entities. Single phrases are perfect examples of concepts. The phrase "apple" refers to that exact type of thinking, fruit. Proposals are statements about the world and can be depicted with a sentence. The sentence "Mary has brown hair" is a proposition that is made up of concepts. Rules are yet some other framework of representation that can specify the relationship between the proposals. For example, the rule "if it's raining, I'll deliver my umbrella," makes the second prelude accidental the first. There are also analog representations; an analogy helps us to compare two comparable conditions. We will talk about all four.

At the end of this, there is a large representation of these elements in the depth section.

There are four outstanding elements of any representation (Hartshorne, Weiss, and Burke, 1931–1958). First, a "representation carrier," such as a human or a computer, must recognize a representation. Second, there should be an illustration content - This means that it is for one or more items. Element or matters in the external world that depictions stand for are referred to as references. An illustration should also be "grounded."

Representation and its context come to be related. Fourth, a representation must be interpreted through some interpreter; both illustrations will tell him, me, or any other person. These and other representational signs. It's symbolic of the fact that representation stands for the potential of something else. We're all familiar with symbols. For example, we understand that the "$" symbol is used for money. The image itself is not real money, but instead, there is a surrogate that refers to it, which is real money. In the case of intellectual representation, we say that there is some kind of symbolic existence "that stands for real money. 1.1 indicates some aspects of the mental representation of wealth.

Mental representations can stand for many different.

There are types of cases not using limited conceptual ideas like simple money." Research indicates that there are more complex mental representations that can stand for rules, e.g., understanding of car and force. Simulations can help us solve certain problems or be consciously equal (Thagaard, 2000). For more specifics, discussion sees an in-depth section of these various forms of intellectual representation. Human mental representations, especially linguistics, are called for Semantics, that is, their meaning. What this means and how important its portrayal can be is a matter of debate. According to one view, the meaning of a representation is derived from the relation between the representation and what it is about that word. This relationship is described intentionally. Intention means "directed at one

Object. "Mental status and opportunities are intentional. They refer to something real, anything in the world. If you think of your brother, then thoughts are directed towards your brother, not your sister, a cloud, or

Something different. The intent is seen to have at least two properties. The first is egalitarianism or the similarity of structure between a representation and its context. This similarity capability can map different components of an illustration

Reference to it. Analog visible images, which are mentioned below, are appropriate examples. This property is considered because they maintain the spatial features of reference. For example, a visual photo of a cruise ship must have increased horizontally and vertically because these are boats they are tall. Researcher Stephen Koslin has shown this. It takes longer to "scan" a visual image where the distance between the points of the object in one dimension is large and incredibly short during such a short time. The section on visual imagery includes the strategies and results of this test and others that exhibit isomorphic characteristics of images.

The second feature of intention has to do with the relationship between input and output for the world. It must be a deliberate portrayal that can be brought up using references or things related to it. Consequently, activation of representation (i.e., questioning it) must be due to behavior or gait

One way or the other is connected to the context. For example, if your pal Sally has informed you an image of a cruise he took in the Caribbean finals in December

The cruise ship will likely be in mind. After that, you can ask him if the food onboard was once good. Sally, on the cruise, was mentioned for excitement.

Digital representation

In a digital representation, also known as a symbolic representation from time to time, data are recorded in discrete ways with set values. A digital clock for the example represents time judiciously (see 1.2). It shows the difference.

The number for every hour, minute or year. There are amazing benefits to digital representations. The values are specified exactly. Symbols used in digital representation, such as numbers, can be driven by more general sets of processes than analog structures. In mathematics, a wide range of operators, including segmentation or squaring, can be used for digital number representations. The effects of these operations are new numbers that can be transformed by additional operations.

Language can serve as an example of a digital mental representation, and

factual oral theories appear to be machines of human symbolic representation that are most commonly used. The basic factors of written language are letters.

These are discrete symbols that are mixed according to a set of rules. The combination or word means combined into itself, other higher-order units, sentences with additional semantic content. With the help of rules in which these word elements are mixed and modified in the language known as syntax. The syntax constitutes a set of permissible operations on the word elements. It is the elements themselves that are mental representations. They are additionally constrained for use in problem-solving. Visual images are excellent examples of analog mental representations. Cognitive psychology researchers have performed a vista experiment. We strongly recommend that we represent the visual record in an analog fashion.

Stop studying for a moment and close your eyes. Imagine a palm photo

Trees on a sunny beach. Can you see the sample on the bark? about what

Coconut? The images capture many of the same houses as their reference, such as the distance between the corresponding units of digits. Different types of changes that can be made on pixels are additional types of modifications.

Physical objects pass through the outer observatory. It contains rotation, translation, and reflection. In the field of cognitive imagination, we discuss the nature of visual snapshot experiments that reveal the operations types that can be performed on them.

Dual coding hypothesisCollective use of both digital/symbolic and photo representation has been referred to as a double-code hypothesis (Pavio, 1971). Alan pivio assumes that multiple views can be represented to either of these two types. This is particularly real for a precise concrete concept, such as "elephant," for which we can structure a visible picture or averbal representation.

However, there are some standards for which a symbolic code is more visible Sui. Think of "justice." This is the essence, and although we can attach an image to it, as in a court building, no one is indicted and special image detection.

Evidence in support of dual-code theory comes from studies that have better remembered for phrases representing concrete concepts as a comparison for words representing summary ideas (Pavio, 1971). According to Patio, the reason for this is that two codes are better than one. Let's guess a topic memory test includes the phrase "elephant" and two varieties.

To know this. If he forgot to remember a code later, he should still be able to get the right of entry and retrieve the other. in this matter, the image of an elephant can also come into consideration even if its symbolic word depiction has faded.

Proposal representative

Apart from motion representation, representation has the third-largest class Symbolic and Fantasy Codes (Pileshin, 1973). According to the proposal

Hypotheses, mental representations are like abstract sentences Structures. The propositions are true by taking the relationship between the concepts. For example, the sentence "Mary seemed to John" is of a type.

The relationship between Mary and John, and that relationship can be translated as a verbal, symbolic code, as in the actual shape of a sentence, or

A photo code.

The proposal is believed to be in a dark layout that is neither visible nor Oral. This structure can be described as a logical relationship between constituent components and is represented with the help of a predicate calculation. A prophetic stone

is a general machine of common sense that accurately expresses a large range of reasoning and methods of reasoning. The proposal appeared in "Mary John."

It can be represented with the help of a predicate calculus such as:

[Relationship between elements] ([subject element], [object element])

Where "Mary" is the issue element, "John" is the object element, and "see" is the relationship between the elements. What is satisfactory about a predicate calculus is that it holds the critical, logical structure of a complex idea independent of its actual elements. Any range of topics, objects, and relationships can be inserted into the summary layout of a proposal. A proposal that, as a consequence, is believed to seize the core of a complex idea. This core

Meaning, when translated back into a symbolic or visible code, can occur

Expressed in many ways. For example, the sentence "Mary appeared there are two alternative verbal codes for "John" and "John was once considered Mary's way"

Same offer. Likewise, one can shape many visible images

Although a predicate calculation is a good way of expressing a proposition.

This does not mean that the proposal considers this layout in our minds at all.

It is unclear exactly how the proposals are mentally instantaneous because they are concise and can specify all of the viable relationships between concepts.

To sum up this section, mental representations are powerful. They allow the beginning of an inner world we can think of. A by-product of these ideas enables us to be taken into confidence and interact effectively with the environment. Instead of knocking and making mistakes in the world or taking the risk, we can use representations to sketch and appropriately take action.

Also, the formal implementation of representations in a set of symbols, as we hypothesize in intellectual pix or language, allows us to speak our thoughts to others. This, in turn, provides more complex and upward thrust

Adaptive varieties of social collaboration.

As mentioned earlier, representation is entirely the first main element.

Cognitive science's view of intellectual processes is represented by themselves.

Except for something, little use can be made with them. Got the impression of

Money doesn't do much for us, except we know how to calculate a tip or a can.

Return the correct amount of business to someone. In cognitive science

See, thinking computes representations. It is, therefore, important to identify how and why these mental systems operate.

Any given statistics procedure can be described at various extraordinary levels.

According to the three-stage hypothesis, intellectual or synthetic information processing events can be evaluated to at least three exceptional degrees (Marr, 1982). The highest or most summative degree of evaluation is the computational level. On this

level, one included with two functions. The first is a glaring specification of what the problem is. Taking trouble, because it can happen in the beginning, in one obscure method perhaps, and by breaking it into its key materials or parts, can bring about this clarity. It can accurately describe trouble in such a way that perturbation can be investigated using formal methods. For example, Questioning: What is this problem required? 2d venture on computational degree faces a purpose for the process. 2D assignments include asking: Why is it like this here in the first place? There is thinking of adaptability inherent in this assessment - the idea that human mental techniques have been learned or advanced to enable a human organism to face trouble. It is the primary explanatory points used in evolutionary approaches. We describe one causative reason for the number of cognitive approaches and their development in chapters devoted to that view.

Bringing down a level of abstraction, we can inquire about the real way an information system is done. To do this, we want an algorithm, a formal technique, or a machine that works on informational representations. It is essential to note that algorithms can be executed regardless of the meaning of the representation; Algorithms operate on a form, no longer meaning the symbols they replace. One way algorithms work is that they are "functions" used to manipulate and make alternative representations. Algorithms are formal; this and how an accurate step makes modifications to the data being acted upon. Mathematical formulas are a reasonable example of an algorithm. Specifies a formula on how to change the placed at the algorithmic level, once known as the programming level. This is equivalent to asking the question: Which information-processing steps are being used to solve the problem? If to draw an analogy with computers, the algorithm phase is like software, because software programs have guidelines for processing data.

The most accurate and concrete type of description has been prepared.

Implementation level. Here we ask: What is a statistics processor? What type of physical or physical changes occurs in the processing of information? This step is referred to as hardware-level since, in computer parlance, the hardware is the physical "stuff" Made of computers; it will consist of many parts - a monitor, hard drive, keyboard, and mouse. On a small scale, computers are hardware wave of electrons through the circuit and even the circuit. Hardware in Human or animal cognition is genius and, on a small scale, neurons and activities of those neurons.

Or the formal level of analysis? Now why not just map physical processes agree of implementation on the computational description of the problem, or

Alternatively, on the behavior or gait of the organism or device? This seems simple, and we don't want in for data and representational thinking. The reason is that the algorithmic phase tells us how a unique system performs a computation. Not all computational structures solve a problem like this. Computers and humans can each add extra, though defense, in quite a different fashion. This is true at the implementation level explicitly, but formalizing the notion reveals much about the alternative problem-solving approach. This additionally gives us insight into how these structures can compute alternatives to other novel perturbations that we probably would not understand.

This division of information-processing opportunities analysis into three

This level has been criticized as being essentially simplified, given that each level of the degrees can be divided equally (Churchland, Koch, and Sejnowski, 1990). 1.4 depicts a possible enterprise of several structural levels of evaluation in the system concerned. Starting at the top, we can consider

Talent as an organizational unit; Brain regions correspond to another step down in the organizational unit spatial scale; And then neural networks,

Individual neurons, and so on. Similarly, we may want to divide the steps of the algorithm stress into specific sub-stages and sub-problems. To include all

This is no longer evident on how to map one level of analysis to another, can also specify exactly how the algorithm performs. The damage is done with the right place or how it identifies the nerve system.

Classical and connectionist view of computational before we finish our dialogue of calculation, it is necessary to distinguish between two different conceptions of what it is. So far, we are talking about calculations based on formal systems assumptions. This PC is a formal image manipulator. Let's ruin this definition in its issue parts below. A system is formal if it is syntax or rule set. Language and arithmetic policies are formal systems because they specify what accept adjustments can be made to the symbols. Formal systems additionally handle the content of these representations objectively. In different words, a process can be applied to an image regardless of its meaning or meaning content. As we have already seen, a symbol can rely on a wide range of forms and forms of representation.

Manipulations

The meaning here is that calculation is a powerful, contiguous method.

Types of computing devices and they take some time to happen, i.e., they do occur as well.

But it is no longer only thought what computation is. The traps approach to computation differs from the classical formal systems approach to a large extent in cognitive science. In the classical approach, information is represented locally, in the shape of symbols. The connective view has knowledge presented as a sample of activation or weight that is distributed throughout a network. Processing fashion is also distinctive in every approach. Classical

The viewing process is going on in discrete stages, whereas in connectionism, processing takes place in parallel through simultaneous activation of nodes. Some cognitive scientists minimize these differences, reasoning that information processing occurs in both structures and may be triangular level speculation equally applicable to each (Dawson, 1998). We examine and distinguish classical and connectionist views on the establishment of the Network Strategy. An interdisciplinary perspective is an ancient myth about five blind people who stumble upon an elephant (see 1.5). Not understanding what it is, they begin to experience the animal. A man thinks

Only the elephant's trunk and thinks that he is feeling a big carrot. Another man, feeling ears, believing the object is a big fan. 0.33 trunk and feel

Declares that it is a pestle, while only touching the fourth leg believes that it is a Mortar. The fifth man touches the tail, but has another opinion: he believes in rope making. All five people are wrong in their conclusions because each has thoroughly examined one digit of the elephant. If five people were found together and shared their findings, what they could be pieced together without further problems

What kind of creature was it? This story serves as a pleasant metaphor for the cognitive sciences. We can think like elephants and blind people

Researchers in specific disciplines in cognitive science. Each character discipline can also make excellent progress in perception in terms of its unique concern, but, If it cannot evaluate its results for other related subjects, it may additionally be exempt

The true nature of appreciation is being investigated. The key, then, is to think of something as mysterious and complex as the mind as there are communication and collaboration between disciplines. What does it mean when someone talks about cognitive science - not the sum of every genre approach, however their union. Recent years have considered this collaboration an amplification. Many core universities have straddled interdisciplinary cognitive science centers, where researchers in many fields, such as philosophy,

Neuroscience and cognitive psychology are always motivated to work collectively on problems. Each place can contribute to its special power under study. Philosophers can present big questions and hypotheses,

Neuroscientists can measure physical performance and brain activity,

While cognitive psychologists can pick up diagrams and experiments. A fruitful synergy is formed after an interchange of results and ideas.

Across these disciplines, accelerate growth with an appreciation for finding solutions insight into trouble and yield other search questions.

We have indicated some extraordinary processes in cognitive science.

Because this book is going to explain every method and its most important principle contribution, it is worth describing each of the phrases from its perspective,

History and methodology. We will also present a brief description of the following sections.

A preview of the issues addressed by each approach.

Philosophical view

Philosophy is the oldest of all subjects in cognitive science. It shows

how its roots returned to the historical Greeks. A lot of recorded philosophers have been lively at some points in human history, attempting to prepare and answer basic questions about the universe. This approach is definitely free for anyone to know. Honestly, important questions on anything from nature to honesty

Existence for the acquisition of knowledge for politics, morality, and beauty. Thinking philosophers focused their attention on related problems in particular

Nature and characteristics of the mind. They would probably ask questions like this: What is mind? How do we recognize things? How is intellectual knowledge organized?

The major method of philosophical inquiry is logical, both preventive and inductive. Deductive logic emphasizes the utility of the rules of logic to make a statement about the world. Given the initial set of given statements true, philosophers can derive other statements that must be logically correct. For example, if the declaration "college students learn about three hours every night, "It is true, and the statement "Mary is a college student" is true, we can then conclude that "Mary will study for three hours every night." In inductive reasoning, they observe unique cases in the world, have common similarities between them, and conclude.

Free deterministic debate tells us whether our moves can happen anytime fully considered and /or anticipated beforehand. Gain knowledge

The problem is how we know things. Is an information product through one's genetic endowment or interaction with its

atmosphere? How each of these factors contributes to any given

mental capacity? We also seem to be one of the most captivating and esoteric

The secret of mind, of consciousness. What is consciousness? Are we really conscious?

Psychological approach

Psychology is a fictional youth discipline compared to philosophy.

However, it is considered old, especially when it is compared to someone and more current learners for the cognitive science scene, for example, artificial genius and robotics. Psychology arose in the late nineteenth century, and first was self-discipline in which scientific techniques were specifically applied and find out about intellectual events. Early psychologists established experimental laboratories that allowed them to list intellectual ideas and see a variety of intellectual abilities, such as imaginative and presenter and memory. Psychologists apply scientific methods for every thought and practice. That is, they're intellectual phenomena as ideas are no longer understood but in addition to these

external behaviors that can push upward inside these events. There is a way to have valid knowledge about the scientific technology

world. It starts with a hypothesis or thought about how the world works and then designs a test to see if the speculation has validity. In one of the experiments that make observations below a set of truly controlled conditions. The resulting data then guide or fail to aid both hypotheses. This process is employed within psychology and cognitive science

Additionally, the general is described at the beginning of Chapter 3. The field of psychology is wide and includes many sub-disciplines, Each with its unique theoretical bent. There is an extraordinary take of every self-discipline that is thought through. Early psychologists, that is, autocrats and structuralists saw this idea as a look at the tube

Chemical reactions occurred between intellectual factors. Conversely, functionalism no longer looked at the idea according to its constituent parts, but according to an imaginative and presenter of thinking as the creation of the parts, however, insisted that it was a combination and intersection of the parts, emphasizing upward to give new emphasis, which was important. Psychoanalytic psychology imagines thought as a collection. Competitive entities, while pragmatism, sees it as a machine that exhibits stimuli

on behave.

Cognitive approach

A new shape of psychology that began in the nineteen sixties came on the scene. Known for cognitive psychology, it came in part as a protest emphasis on understanding internal intellectual operations.

He adopted the laptop as a metaphor of the mind and described intellectual functioning in phrases of representation and computation. He believed that the mind, like a computer, can be understood in phrases of data processing. The cognitive approach was additionally capable of explaining such phenomena. Language acquisition, for which behaviorists no longer had the right accounts. Around the same time, new applied sciences allowed better shape mental effort was being developed. It gave a boost

The practitioner's emphasis on external observable behavior towards internal scientific work on the internal activity, as, for the first time, it can be found with sensible precision.

There is a concept of modularity inherent in the cognitive strategy. Modules are functionally independent mental devices that receive input from various modules,

Perform a precise processing task, and bypass the results of their calculations additional modules. The modular strategy may have an impact on the use of a process model, or a wafer diagram is seen. These depict a governmental exercise using packing containers and arrows, where boxes depict modules and arrows the flow of facts between them. Methods used in the approaches are experimental approach and computational modeling.

Computational modeling involves a formal (usual software) implementation of a proposed cognitive process. Researchers can remodel the system to simulate how the technique can work in a human brain. They can then change several parameters of the model or change its structure to achieve results as closely as feasible to achieve in human experiments. This use of comparison and modeling with experimental data is a special feature of cognitive psychology and is also used.

Artificial brain and community perspectives.

Cognitive psychologists have studied many intellectual processes.

These patterns include recognition, attention, memory, imagination, and problems. Theoretical accounts and processing fashions are given for each of these

In chapter 4 and 5, language is within the realm of cognitive psychology, although the approach to language is also multi-disciplinary because it is described as cognitive psychology in Chapter 9 separately.

Neuroscience approach

Brain physiology and anatomy have been studied for some time.

On the other hand, in recent times, there has been tremendous progress in our understanding. In the brain, especially in phrases about how neuronal processes can occur

Cognitive phenomena. General discovery about common sense and the endocrine systems known as neuroscience attempts to explain cognitive techniques in the context of the underlying brainstem and is considered to be cognitive neuroscience. Neuroscience presents, first and foremost, a description of mental phenomena

At the implementation level. It tries to describe the biological "hardware" on which intellectual above, there are several stages of the scale when it comes to describing the brain, and

It's not usually clear to what degree confirms any given excellent explanation of the cognitive process. However, neuroscientists investigate at each of these levels they learn about the character neurons and neuron-to-neuron cell phone biology

Synaptic transmission, pastime patterns in neighborhood cell populations, and interrelationship of large geniuses regions. One of the motives of many recent developments in neuroscience is, again,

the development of new technologies. A wide range of neuroscientists measures the overall performance of the genius at working machines. It contains Positron emission tomography (PET) scanners, computerized axial tomography (CAT) scanners, and magnetic resonance imaging (MRI) machines. Contributors have a cognitive function in studies using these tools;

The brain entertainment that is concurrent with the performance of the task has been recorded. For example, a participant may also be asked to create a visual picture of a word shown on a laptop's screen. Researchers may determine which aspects of the brain have become involved for a period of imagination and in what order. Neuroscientists use other methods as well. They study brain-damaged victims and effects.

Wounds in laboratory animals use single- and multiple-cell recordings techniques.

Network approach

The community approach derives at least partially from neuroscience. In this perspective, the mind is considered as a series of character computing units. Units are connected and affect each other's activity simultaneously through the connection. Although every gadget is supposed to perform relatively easy computation, for example, a neuron either firing or not.

Firing can provide upward just for connectivity, representational, and computational complexity of devices. Which outlines of the network's approach has two parts? First,

the development of synthetic neural networks is involved. Most artificial nerve networks are computer software simulations designed to be emulated.

The way authentic talent networks operate, or the functioning of neural telephone populations. Artificial neural networks that can perform arithmetic learn concepts, and loud scrutiny now exist. A wide variety of network architectures has been developed over the last thirty years. The second phase of the network is also theoretical and focuses on the representation of knowledge - how meaningful facts can be mentally as well as coded and processed. In semantic networks, the nodes standing for the concept are the interconnection of one node in such a way that activation of one node causes activation of different related nodes.

Earth Network is built.

Remember how conceptual statistics are equipped and memorized in memory. They are often used to make predictions and provide explanations for facts derived from experiments.

Evolutionary approach

Proposed the principle of the natural decision using Charles Darwin in 1859

Our way of thinking about biology revolutionized. Natural determination holds

Adaptive features allow animals that have them to live and be near these aspects for generations to come. In this scene, the surroundings are considered

Choosing one of these traits that serve a purpose.

The evolutionary method used to provide an explanation for the phenomenon of biology. The topic of evolutionary psychology applies the selection concept to human mental processes. This attempts to clarify our ancestors and how the decision operates those forces that gave rise to the cognitive constructions that we now have. Evolutionary Psychologists also take a modular approach to the mind. In this case, the modules are "favored" analogs of cognitive abilities that were used by ancestors successful in solving positive problems. Evolutionary theories have been proposed to explain the experimental results in varying capacities—classification from memory, logical and probabilistic reasoning, cognitive variation between language and sexes.

There is a version of this subject called Evolutionary Computing, in which the rules of

Evolution is applied to create profit computer algorithms. A crime of this form of computing is artificial life. These are software program simulations that mimic the Biological Ecosystem. In addition to this is neural Darwinism, which uses evolution to explain the formation of neural circuits. See Chapter 8 for more on these. Linguistic approach linguistics is a place that focuses entirely on the field of language.

It is concerned with all questions related to language ability, such as: What is language: Hindi? How do we get the language? Which components of talent use language? As we have seen, neuroscience.

Because many separate disciplines become more united through the subject

By approach or method. Part of the problem with reading language is the fact that language is so complicated. Much has been done to understand its nature. This work looks at the properties of all languages, factors of language, and how to view the elements that are used at some level in communication. Foci of linguistic inquiry corps on other oriental language usages, language acquisition, and reduction in language acquisition

Early sensory deprivation or intelligence damage, the relationship between language and thought, and the development of speech awareness systems. Linguistics, perhaps more than any of the different approaches discussed here, adopts a very liberal methodological approach. Language researchers hire experiments and computer models, learn about brain-damaged patients, how to tune language

capability adjustment at some point in development, and evaluation of multiple languages.

Artificial intelligence approach researchers are creating gadgets that try to imitate humans and animal ceremonies for many centuries. But it's completely in the past computer scientists that have seriously attempted to manufacture devices that mimic complex thought processes. This field is now known as Synthetic Intelligence (AI). Researchers in AI are concerned with carrying out duties in computer systems where human intelligence is required as they build programs to do a variety of things that require complex reasoning. AI packages have been developed that can diagnose scientific disorders, use language, and play chess.

For the second time, AI gives us insight into the function of human intelligence operations.

Designing a PC program that can see an object often proves useful in demonstrating how we can do similar tasks ourselves. A more exciting result of AI lookup is that sooner or later we can create a synthetic character that will possess all or several points that we

consider unique human beings, such as consciousness, decision-making ability, and so on.

This is an evaluation and subsequent development of laptop algorithms and their tests, empirical facts or performance standards.

The modification that constitutes the methodology of AI perspective. Not all of the computer packages, however, are similar. Researchers have made a detailed working approach. An initial effort to get computer systems to the joint application of logical policies to propositional statements. Later, on expert systems, scripts, and fuzzy common sense processes, among others, were used. Chapters 10 and 11 provide precise descriptions of these techniques.

Robotics approach, finally, we show on consideration on robotics. Robotics can be considered a family affair

AI and has recently considered the scene as a formal discipline. While AI people build units that "think," robotics researchers make

machines that must also "work" build investigative autonomy in this area

Or semi-autonomous mechanical gadgets that are designed to perform

a physical mission in a real-world environment. Examples of cases of robots

Currently, a disorganized room may consist of welding or navigating around, manipulating parts on an assembly line, and diffusing bombs. There is a great deal to contribute to and do in cognitive science in robotics strategy

Principles of mind. Like humans and animals, robots have to demonstrate successful goal-oriented behavior in complex, changing, and uncertain environmental conditions. As a result, robotics helps us to think.

In Chapter 12, we define special paradigms in robotics. Some of these approaches are fundamentally different from each other. Provides a hierarchical pattern "top-down" perspective according to which a robot is programmed with

Information about the world. The robot then uses this model or internal illustration to describe its functions. The reactive pattern, on the other hand, is

Down up." Robots using this structure reply in an easy way.

Environmental stimuli: They respond consciously to a stimulating penetration and have mental representation-Digital, analog, and proposal - each has its characteristics, and we gave an example of each. However, the history of research into cognition suggests that there are also serious forms of mental representation, Paul Thagard.

Mind: Introduction to Cognitive Science (2000) proposes four. These are propositions, concepts, rules, and analogies. Although some of these have already happened They are alluded to and described elsewhere in the book. They are central to many views in cognitive science. So it is useful for some of their layout key features here.

An idea is possibly the most fundamental form of mental representation. It is a concept that represents the cases that we have grouped.

The "chair" no longer refers to a particular chair, such as the one you're sitting on.

Now, but much more than that is customary. It refers to all possible chairs that do not matter their colors, shapes, and sizes. Concepts do not want to refer to solid objects. They can stand for summary ideas, for example, "justice" or "love." The concepts may be related to each other in complex ways. They can relate hierarchically, one level of thinking is for everyone at one level of the organization

embers of the classification just below it. "Golden retrievers" belong to the class of "dogs," which belongs to the category of "animals" in Flip. We discuss the hierarchical effigy of perception presentation in the community approach.

On the question of whether standards are a comfort or real In the Philosophical Method. The Synthetic Genius outlines the use of buildings known as representing conceptual knowledge.

A motion is a statement or a claim usually presented in the shape of a simple sentence. An important function of a proposal is that it can be proved right or wrong. For example, the declaration "Moon is made of cheese."

It is grammatically correct and may also be characterized by a belief that some people hold, but it is a false statement. We can apply formal common-sense rules to the proposal to determine the validity of these proposals. It is referred to as a logical inference Rationale. Impotence consists of three propositions. The first two are the campus, and the final one is the conclusion. Take the following impotence: All men love football.

John is a man. John likes football.

Obviously, the conclusion can be false if both have their premises.

It is wrong if it is no longer authentic that all men love football; it probably would not be fair. John loves football, even though he is a man. If John is no longer a man, he can either.

I don't like football anymore; everybody likes it. This type of logic similar to the reasoned

You may have also seen that propositions are representations that contain concepts. The offer includes "all men like football."

Concepts "Men" and "Football." Proposals are more sophisticated representations than ideas, because they are specific relationships, from time to time complex, between concepts. The rules of common sense are considered great as computational methods that can be used to determine propositions of their validity. However, members of the logical family may also be themselves in the middle of the proposal

A different type of representation is considered.

Evolutionary approach

This gives an interesting account of logical reasoning, which for many people, is difficult and less complicated than positive situations. Logic is no longer the only machine to operate on offer.

The rules also do this. A manufacturing rule form has a conditional declaration: "If x is, then y," places x and y are propositions. The "if" phase of the rule is called Bet. The "then" phase is known as the action. If the proposition contained in condition (x) is true, then the speed extended through another proposition (y) should be done according to the rule. Following rules help us stress our cars:

If the light is red, step on the brake. If the light is green, step on the accelerator. Note that, in the first rule, the two propositions are "light red" and "step on the break." We can create more complex guidelines through linking proposals with if the light is crimson or light yellow, then step on the break. If mildly inexperienced and no one is in the crosswalk, then step on the accelerator. The "or" hyperlinking the two propositions in the first section of the rule specifies that if both propositions are true, action must be taken. If one "And" combines these two propositions, the rule specifies that both must be true before the proposal arrived.

The rules reveal just what knowledge is. We usually think of knowledge as factual. A proposition such as "candy is sweet," if validated, provides accurate information. The proposal is an example of declared knowledge. Declarative information is used to represent facts.

It tells us what is using oral communication. Procedural Knowledge, on the other hand, represents skill. It tells us how to do something that is validated using the action. If we say that it was fought during World War II

We ski on the slope of an icy mountain in winter, we have recognized it.

We have a unique skill. As a result of this, the information system must have some way of representing the movements in the system if they are to help organisms or computer to perform those tasks. Rules are a way of expressing procedural knowledge. We talk about two cognitive rules-based systems, Atomic Components of Thought (ACT) and SOAR Models in Cognitive Method Chapters.

However, as indicated below, another specific type of mental representation analogy may also be labeled as the size of the argument.

Thinking anciently involves using an acquaintance with an ancient situation for a new position. Suppose you never rid the train before, though, took buses at different times. You should use your perception of riding on the bus to learn how to ride a train. Information that you have already applied

The pass that applies to both conditions will allow you to complete this. Based on prior experience, you would have understood that you have to do it first. Set a schedule, perhaps decide between a hierarchical and adjoining service, buy a ticket, wait in line, board, stow your luggage, locate a seat, and so on. Simulations are a beneficial form of representation because they allow us to normalize our learning. Every situation in existence is not new. We can do to apply what we have already felt similar situations apply to discover everything. Multiple models of analog logic has been proposed (Forbes, Gentner& Law, 1995; Holyoak& Thagard,1995). We outline some features of analog reasoning in the minds of this area. If you want to try the remedy, you can change it now

Corresponding logic problem.

Analog software artificial genius is referred to as a case-based argument and is subsequently described in the artificial Genius Chapter.

Suppose you are a doctor facing a patient who has a malignant tumor in his stomach. It is impossible to act on the patient, although excluding the tumor destroyed, the patient will die. A beam of sufficiently high intensity can destroy the tumor.

Unfortunately, the full tissue at this depth

The rays will be destroyed on the way to the tumor. On essays of intensity are innocuous to the full tissue, although now there will be no effect on the tumor. How can rays be used to destroy tumors besides injuring the health issue? And here is another story, called a "familiar and fort problem."

please read it. Does it help you solve the tumor problem how? The fort was once dominated by a small strong fortress in the United States that is located in the center of the village, surrounded by farms and villages. Several roads pass through the fort through the countryside. A revolt universal pledge capture the fort. Standard already knew that his entire navy would attack through capture the fort. He gathered his army to launch a full-scale direct attack. However, the customary then discovered

The dictator had mines on every road given the need for a dictator; small bodies of men may want to pass over them safely. Take your troops and people away from the fort. However, any big force will explode in the mines. This will not only blow the road, but it will also destroy many neighboring villages. It seemed impossible to catch out. However, Famous devised a simple plan. He split his army in small businesses and sent each crew to the head of a particular road when everyone was ready, they signaled, and each crew drove down a specific road. Each team continued down its street to the fort at the same time, and they reached the fort together. Like this, you may have seen many similarities between these two stories. The tumor is compared to the fort. Rays were to be used to damage the tumor-like soldiers that are desperate to penetrate the fort. Healthy tissue in the first story, another, villages can be compared. Given these similarities, you may have a similar answer applied, which the rebels commonly used to cause trouble.

Eradication of tumors. Like a solution to divide and send troops, soldiers built separate roads to convert to the citadel, the solution of the tumor

The problem forces a high-intensity beam to be divided into a pair of low-intensity rays. Focuse them on the tumor at specific angles. In this way, the rays

Healthy tissue surrounding. Gick and Holyoak (1980) found that only 10% of participants who know about them should effectively overcome the tumor problem in the being provided with their general and fort story. A full 75% of the participants were effectively resolved when they were supplied with the story.

Solve a situation or problem. Tumor problems in this example

also, hire an existing analog that is derived from another learned position. This is a traditional and fort problem. The analogy is systematic relationship similarities noted above, and between these two analogs are included. The majority of analog model reasoning suggests that the method has four levels of methodology analog logic. The first goal is understanding the problem. Another is recalling a similar supply problem for which a solution is already known. After this, source and target issues are compared. This is equivalent to mapping similarities in their corresponding structures. As a final step, there is a supply problem optimized to produce solutions to the target problem.

Ideas for food: discussion questions

1. Many metaphors have been proposed to think about the mind.

From water pumps to smartphone systems. Can a corporate workplace serve as a metaphor for the mind? Why or why not?

2. Solid concepts like "snakes" can be represented separately

Abstract concepts such as "democracy"? What kind of concept lends itself

More without difficulty for analog representation? Why?

3. Describe how a handheld pocket calculator performs the division at the computational, algorithm, and implementation levels of analysis.

4. Concepts, propositions, Images rules, and analogies are all types of mental representation. Can you think of different examples?

5. Think of an opportunity in everyday existence in which you used analog logic.

Describe the problems of the goal and the source as a good deal as possible, and similarities between them.

Although functional science has been the dominant approach in cognitive science

In the 1970s, it is not without its shortcomings (Maloney, 1999). Remember that the principle of functionalism is that brains that are no longer solely based on brains can exist. They can exist in cases such as computer systems as long as physical substrates allow for enforced computation to those objects. Critics argue that,

While it is plausible that the brain can exist in the absence of the brain, it does. This is not praiseworthy. There is no empirical evidence of modern times to justify this claim. We have not yet seen anything intellectual in the absence of a brain. Other than this, some have argued that failure to discover the idea with the physical form.

The reason to distance oneself from the concept of mind should be considered - rather than has a unique reputation as a purposeful type. An additional problem with functionality is that it cannot calculate mental state feeling or skilled personality - a phenomenon considered qualitative (quale, singular). Examples of qualia include subjective riding; it is like being "hungry," "angry," or looking at the shadow "red." These types of experiences do not seem to be replicated as inherently functional.

A computing device can be programmed to make the color "red," even mimicking the same human practical process, but this machine should not be so far a man or woman likes to see the purple he has.

What's more, two humans experience the same consciousness.

Now do not travel it equally evenly. A quantity of experiments is shown in the case of shadow perception. Searching for participants, the same color would describe it in another way (Chapanis, 1965). If asked to hint on a pure spectrum that looks pure green, a character can additionally select

Yellow-green, second blue-green. Although this case functionally operates respective minds as they see that the colorization is almost equal. In this case, the neurophysiological operations behind color comprehension are similar in individuals. Think about all the options in your life for a minute. some for examples, it is important to know which college or college is for whom

Join, or possibly identify whether or not to pursue a romantic relationship. Others may also be less necessary, for example, to get pork with or without fried rice or Sichuan foul for lunch. Free will debate

It is about whether these behaviors are under our control. Did you consciously punish one college vs. another for appearing in the exam potential execution and opposition of each, or forces beyond your manipulation pushed you to attend one school over others? This will be the side of free takers debate argues that people independently incite their actions. Those Advocate determinism argues that moves can be defined in the phrases of the initiating motives that precede them, meaning that people are entirely dependent on them as the end result of these causes. We should start with determinism. The idea that all physical opportunities are the cause or decision was taken with the help of the full amount of all previous events.

Our actions, which are physical events, consequently, it should also be determined. Suppose you get up to take a snack from the refrigerator. According to determinism, you rise every other physical tournament (or event) was aided by what it had just been. The match may be belly pang or listening to the commercial restaurant on the radio. Furthermore, the deterministic view is that it was unavoidable

You may have taken this offer because of the preceding event. In other words, no other action can be taken by you in view of nature. Who are you and the set of opportunities that preceded the action? Truthseeker David Hume (1–11–1 con suggests☐) suggests that we imagine a firm universe in the context of the billiard balls. To experience this notion, imagine that, initially, a set of billiard balls is scattered on a and balls capture random positions. We then come with the cue stick and give a knock ball in second. Each shifting ball, in this human effigy of work.

The event that once transpired through the previous match and the reasons for every other flip (see 2.5).

This effigy has various implications. The first, as mentioned earlier, is determinism. Ball A hits B, then Ball B can be brought about entirely through the action of ball A. Of course, it is a Simplified Model. A pair of forces can appear in ball B,

In which case, their combination affects the action of Ball B. The point is that Ball B's speed is determined exactly by the powers it sees. Temporarily preceding moments. A 2D implication is a replication. If we were to return all balls to their authentic positions and strike with the ball A Cue stick again, the same way, then all the balls will move once

In the same way, the 0.33 implication is predictive. If we know the condition of all balls and prerequisites under which the first ball will be hit, as if the force is applied and the cue stick moves, then you can know ahead of time what all the balls on the desk will do.

One reason is that these residues of the machine reveal scientists' perceptions of variable interactions in a managed scientist. The actions of billiard balls on each fraction describe the way the experimental variable works. The hypothesis of influencing every other. Repetition and predictability

The scientific method has two corners. We will talk about its extra-scientific method in psychology. If we translate this deterministic effigy of determinism to human behavior, then we can predict any speed that can begin for a man or woman to miss. His or her life. All you need is an understanding of the device and understanding the forces seen on it. In this case, the gadget is the brain. Those who work on force

Talent may include colic or effects such as radio advertising. The behavior of this system will be the superconductor given by GeniusGetting up, like getting up to get a snack. This notion is no longer a long way that behavioral psychologists proposed in the early twentieth century. In his case, he did not explore the idea or how to identify it operated; That is, he notices the machine itself. They focused instead on understanding the causal relationship between stimulus input from the environment and the resulting behavioral output. Some behaviorists felt the theme could fully predict a person's movements based on the conditioning history of that person, which is the time of rewards or punishment.

Has we've come a long way in his life. We will talk more about behaviorism the following.

Most humans find the uncertainty of the cause of the hedge on billiard ball mannequin.

This is because it turns us into automatons and reacts in considered ways

Forces imposed on us. We would like to believe that we choose ours instead of your own path of action. The free approach treats behavior as stem selection or assignment of a will. Selection is done autonomously and is self

Not under the influence of any previous causal factors. In different words, the will. The only decisive motive is the person's action. Not only the product of any separate cause, and is believed to be its own cause. According to this view, people are not about the powers of their past in this way

Twenty-first-century truth seeker Ayn Rand (1963) has created a unit

The model of work, which, according to Rand, underlies free will. In this model, institutions with unique identities are the cause of actions. An entity is defined as an object in the form of a fairly impartial action. It is an example of a character unit. Movements are decided by a unit not using something

The antiqued factor that acts on the entity, but through the nature of that entity as an alternative. If a unit is a fixed method, it will function perfectly according to that method.

In fact, it is not feasible for a unit, which is vice versa. Alternatively, nature, by creating entities due to verbs, Rand shifts the force of work from the environment and towards individuals.

Rand similarly argued that humans are beings of unsure consciousness.

This is the ability that shows that people are capable of thinking, they must create.

Select to do so. From this account, both thought and decided to think or no longer clauses of human nature. If we decide to think, we consciously control our actions. If we fail to think, we are in our minds

Subconscious allied processes. It is the idea of voluntarily directing someone and considered call focus. Cognitive psychologists refer to this more approach.

Evaluation of free will

One problem with free will is that it violates an important notion of work-cause, which is that there should be a reason for all activities. The reason is that the universe is seen here as a network of dependencies, fully existing until they become real through one triggering event or events. If this is true, how can someone provocation come "out of nowhere"? The strict version of the free will implies that the selection or desire for the act of a man or woman is a purpose that is itself unprovoked. It cannot account for the scientific and causal view of the universe.

Chapter 2.

Explanation of this new psychotherapy technique

Psychiatry is a well-known term used to describe techniques for treating psychiatric disorders and mental distress using oral and psychological techniques. During this process, a trained psychiatrist helps the customer locate the source of specific or well-known problems such as precise intellectual illness or existential stress.

Depending on the method the therapist uses, a wide variety of techniques and strategies can be used. However, almost all types of psychotherapy involve building a therapeutic relationship, establishing and enhancing dialogue, and overcoming challenging thoughts or behaviors.

Psychiatry is increasingly seen as a wonderful profession in itself, although many exceptional types of specialists regularly engage in psychiatry. Such humans include medical psychologists, psychiatrists, counselors, marriage and home therapists, social workers, intellectual health counselors, and psychiatric nurses.

Types of psychotherapy
When many people hear the word psychiatry, they imagine a patient lying on the couch, while a therapist sits down near a chair, laying down thoughts on a yellow notepad. There are literally many methods and practices used in

psychiatry. The specific technique used in each scenario can fluctuate based solely on a range of factors, such as the therapist's coaching and historical past, the client's preferences, and the client's current specific nature of the problem.

Some of the strategies required for psychotherapy include:

Psychoanalytic Medicine: While psychiatry was practiced again at some distance as far back as the time of the ancient Greeks, its formal beginnings were when Sigmund Freud began using the measure of discussion to work with patients. Some of the methods used many times by Freud included transition, dream interpretation, and evaluation of free association. This psychoanalytic strategy involves incorporating the patient's thoughts and trying to unravel past experiences, unconscious thoughts, feelings, and memories that may also influence behavior.

Behavior Therapy: When behaviorism became a major dominant faculty of thought at some point in the early twentieth century, methods such as extravagant types of conditioning in psychiatry began to perform an essential function. Although behaviorism may still not be as effective as it once was, many of its methods are still very popular today. The behavioral measure often uses classical conditioning, operative conditioning, and social buyers are getting to know how to help change complex behavior.

Humanitarian Therapy: The School of Ideology was known as humanist psychology beginning in the 1950s, and influenced psychiatry. Humanist psychologist Carl Rogers has developed a strategy known as customer-centered therapy, focused on showing the therapist an unconditionally wonderful relation. The factors of that approach are widely used today. The humanist approach to psychiatry is geared towards helping people maximize their potential. Such processes emphasize the importance of self-discovery, free will, and self-realization.

Cognitive therapy: The cognitive revolution of the 60s has also had a fundamental impact on the practice of psychiatry, as psychologists began to have a growing number of focal points to influence the conduct and functioning of human perception techniques. The cognitive measure is based on the assumption that the power of our thoughts affects our mental well-being. For example, if you look at the negative components of every situation, you will, in all probability, have an extra-pessimistic attitude and a depressed general mood. The purpose of cognitive therapy is to experience cognitive distortions that give rise to this type of question and replace such ideas with something more realistic and overwhelming. In this way, people can improve their mood and general well-being.

Cognitive-behavioral therapy: Approaches accepted as cognitive-behavioral measures (CBT) are a type of psychotherapeutic treatment that helps patients understand the thoughts and feelings that affect behavior. CBT is commonly used to deal with various problems associated with phobia, depression, addiction, and anxiety. CBT is a type of psychotherapeutic approach that involves cognitive and behavioral techniques to change deviant thoughts and misleading behaviors. This approach emphasizes improving the underlying ideas that contribute to the grief and editing of problematic behaviors resulting from these ideas.

Forms of psychotherapy
Psychotherapy may additionally take several different codecs depending on the style of the physician and the needs of the patient. Some that you will probably stumble upon include:

Personal therapy, which works with a psychiatrist to do one-on-one. Couples therapy, which involves a therapist working with a couple to learn how to improve their relationship.
Family therapy, which enables household dynamics and can involve more than one person within a domestic unit.
Group therapy, which forces a small group of humans to share a common goal. This strategy allows team members to provide and receive help from others, as well as practice new behaviors inside a receptive and supportive group.
Before trying psychotherapy, there are some things to keep in mind.
There are problems or concerns for both the physician and the client. When deciding on a physician, consider whether or not you feel comfortable splitting a private record to the physician. You should also check the doctor's qualifications, including a degree to know the years of experience.
Those providing psychotherapy may preserve many of the same degrees or degrees. Some titles such as "psychiatrist" or " psychologist" are included and raise the exact educational and licensing requirements. Some of the people qualified to conduct psychotherapy include

psychiatrists, psychologists, counselors, licensed social workers, and psychiatric nurses.

When providing services to clients, psychiatrists would consider problems such as informed consent, the privacy of the affected person, and the responsibility to warn. Informed consent involves informing a buyer of all feasible risks and benefits associated with treatment. This includes specifying the specific nature of the treatment, any possible risks, costs, and options.

Because clients talk about issues that are highly non-public and sensitive regularly, psychiatrists have an acute responsibility for protecting a patient's privacy. One example, however, is that psychiatrists have a suitable place to breach the privacy of the affected person if consumers offer a near-term opportunity to themselves or others.

The duty to warn grants the counselor and physician the right to breach confidentiality if the client becomes a threat to another person.

How effective is psychotherapy?
One of the major criticisms involved in opposing psychiatry is that its effect is called. In an early and regularly cited study, psychologist Hans Eysenck found that two-thirds of members improved or recovered on their own within two years, even if they had achieved psychotherapy.

However, in a meta-analysis that looked at 475 different studies, researchers found that psychopaths were once of higher quality in improving clients' psychological well-being. The statistician and psychologist Bruce Wampold, in his book *The Great Psychotherapy Debate*, noted that elements such as the therapist's personality, as well as his belief in the effectiveness of the remedy, played a role in the effects of psychotherapy. Surprisingly, Wampold cautioned that the type and theoretical foundation of the measure no longer affect the outcome.

How to know if you need psychotherapy
While you will probably recognize that psychotherapy can help with life's problems, it can sometimes be difficult to seek help or even understand the time to talk to a professional about it.

An important point is that the sooner you are looking for help, the sooner you will begin the journey of relief. Rather than waiting for your symptoms to get out of control, it would be best if you considered receiving help as soon as you notice that there may be a problem.

Some important signs and symptoms to see a psychiatrist may include:

Trouble is causing great misery or disruption in your life. If you believe the problem you are facing disrupts the number of essential areas of your lifestyle, along with school, work and relationships, then it may be time to look at how psychotherapy can help or not.
You are relying on unhealthy or risky coping mechanisms. If you smoke, drink alcohol, overeat or help others overcome their frustrations, then looking for help can help you discover more healthy and more advisable strategies.
Family and friends are concerned about your well-being. If it has reached a point where other humans are concerned

about your emotional health, then it may be time to see whether psychotherapy can enhance your psychological state.

Nothing you have tried has helped. You've read self-help books and online articles or just tried to ignore the problem, but matters seem to remain the same or even worse. Just understand that you don't have to wait until your issues become so heavy that it seems impossible to copy. Help is available, and the faster you get out, the more you will return to a healthy, happy state of mind over the music.

Selecting a medical technique and physician

If you think you have a problem that you can get help from psychiatry, then your first step may be to discuss your concerns with your principal care physician. Your doctor may decide in advance about any physical ailments that may contribute to your symptoms. If a different reason cannot be found, your medical doctor may also refer you to a mental fitness professional who is certified to diagnose and deal with the signs you are experiencing.

The type of therapy is often influenced by your symptoms and the type of physician you choose. If the doctor suspects you have problems, may require the use of drugs in addition to psychotherapy, she may also refer you to a psychiatrist. A psychiatrist is a clinical health practitioner who can prescribe medications and provide proper training in the treatment of psychiatric conditions.

If your symptoms suggest that you may benefit from some structure of the speaking remedy without the addition of prescription drugs, you may also be referred to a scientific psychologist or counselor.

Referrals from friends and family participants can also be a great way to locate a therapist who can help deal with your concerns. However, psychiatry is both an art and a science. If things no longer work or just don't "click" with your cutting-edge doctor, don't be afraid to look for other experts, until you find someone you can connect with.

As you evaluate a psychiatrist, think about some of the following questions:

- Does the doctor feel professional and qualified?
- Do you feel comfortable sharing your experiences and your feelings?
- Do you like the conversational style of the therapist?
- Are you happy with the extent of your interaction with the doctor?
- Does he or she appear to make you feel what you are feeling?

Many words
Psychotherapy can come in many forms, although all are designed to help overcome psychological issues and lead to a higher life. If you suspect that you are additionally experiencing symptoms of a psychiatric disorder, then think about consulting a skilled and experienced psychiatrist who is qualified to assess, diagnose, and treat such conditions. You can reap the viable benefits of psychotherapy even if you understand that something "off" in your life will probably be enhanced with the help of a mental health professional.

Chapter 3.

How can the ACT approach help you if you suffer from anxiety, depression, anger attacks, and panic?

Everyone feels problems or is fearful from time to time. Anxiety is a normal human response to disturbing situations. But those fears and issues aren't temporary for people with anxiety disorders. Their anxiety persists, and can also worsen over time.

Issues of concern can severely impair a person's ability to function in work, faculty, and social situations. Anxiety can increase a person's relationships with family and friends. Fortunately, however, there are fantastic treatments for anxiety.

Drugs sometimes play a role in treating anxiety disorders. Yet research suggests behavioral treatment, alone or overall with medication, is a tremendously effective therapy for most people with anxiety disorders.

Understanding anxiety
Anxiety disorders are common in both children and adults. According to the National Institutes of Mental Health, about 18 percent of US adults and 25 percent of young people, from the age of thirteen to 18, will experience anxiety.

About 4 percent of adults, and about 6 percent of adolescents, term anxiety disorders as serious.

The major types of anxiety disorders are:

Generalized anxiety illness is through power, fear, or anxious feelings. People with this disorder are afraid of many concerns, such as fitness troubles or finances, and it can also be a common sense that something sinister is about to happen. Symptoms include difficulty concentrating, restlessness, irritability, muscle tension, sleep problems, and usually feeling sideways. Common disorders are characterized by recurrent panic attacks, including symptoms such as sweating, trembling, shortness of breath or a feeling of choking, a rapid coronary heart rate, and feelings of fear. Such attacks are regularly sudden, except for warnings. People who have panic attacks often become fearful when the next episode occurs, which may cause them to alternate or stop their everyday activities.

Phobias are severe fears about positive objects (spiders or snakes, for example) or situations (such as flying in an airplane) that are disturbing or intrusive. Anxiety disorder caused by nervousness is also considered a social phobia. People suffering from this disease are afraid of social situations in which they may have to be embarrassed or judged. They typically spend fearful time in social settings, feel self-conscious in front of others, and worry about being rejected by others. Other frequent signs are having a challenging time making friends, avoiding social situations, demanding days before a social match, and feeling unsteady, sweaty, or nauseous while spending time in a social setting. The persistent-compulsive disease is characterized by persistent, uncontrollable feelings and thoughts (passions) and routines or rituals (compulsions). Some frequent instances force hand washing in response to germs' concerns or to check the work for frequent errors. A natural physical disaster such as severe physical or emotional trauma, post-traumatic stress illness (PTSD) may increase after a serious accident or crime. Symptoms consist of flashbacks of trauma, nightmares, and hallucinatory thoughts that interfere with a person's everyday routine for months or years after an annoying experience.

Seeing a psychologist about anxiety disorders

Although a variety of worrying problems exist, research suggests that most are pushed by similar underlying processes. People with anxiety disorders come from difficulty without being overwhelmed by their emotions, and they react particularly poorly to those unbearable feelings and situations.

Often, humans try to cope with the horrific reactions that concern them by confronting those situations or experiences. Unfortunately, rescue can backfire and certainly feed anxiety.

Psychologists are skilled in diagnosing nervous disorders and directing patients to a more effective approach to staying healthy and coping. A form of psychotherapy known as a cognitive-behavioral measure (CBT) is highly potent in the treatment of nervous disorders. Through CBT, psychologists help victims to discover and manipulate factors that contribute to their anxiety.

Through the cognitive issue of therapy, patients investigate to understand how their thoughts contribute to their anxiety symptoms. By knowing how to trade those thought patterns, they can limit the likelihood and intensity of panic symptoms.

With the behavioral component, patients learn strategies to reduce unwanted behaviors associated with anxiety disorders. Specifically, victims are influenced to do things and situations that cause anxiety (such as being in a public conversation or enclosed space) to know the consequences of their fear (such as losing their train of concept or panic attack) is unlikely.

Psychotherapy for anxiety disorders: what to expect
Psychotherapy is a collaborative process in which patients and psychologists work together to identify specific concerns and improve concrete competencies and techniques to combat anxiety. Patients who practice their new abilities may consider staying out of periods to manipulate anxiety in a situation that will likely make them uncomfortable. However, psychologists do not push victims into such incidents until they are convinced that they have the skills they need to overcome their fears better.

Psychologists use other methods to treat anxiety issues other than CBT each time. Group psychotherapy usually involves countless humans who all have anxiety disorders, and It can be fantastic to apply to patients with treatment and support for anxiety. Family psychotherapy can help household contributors understand their loved one's anxiety, and help them learn ways to engage that do not reinforce anxious habits. Family therapy may be particularly useful for youth and adolescents with anxiety disorders.

Anxiety disorders are very treatable. Most victims who suffer from anxiety are in a position to limit or overcome the signs and symptoms after several (or fewer) months of psychotherapy, and many victims follow enchantment after only a few sessions.

Chapter 4.

Because the very nature of human language can cause suffering.

Communication and perception among people, ages and cultures. From the beginning, humans have been involved in social contexts of unique stages of complexity, and they live because it is each putting in their labor and leisure, even when they consider themselves isolated. Endless invisible threads link them to the existence of society. The whole essence of man, consisting of his consciousness, is communicable with the help of his nature. This ability defines the essence of cognition and its vehicles, man or woman, and society. People are affected continuously by communication environments. They are eager to say something, to study or to teach, to show or prove, to agree or reject, to console, to transplant, to show love, and so on. Communication arose, and society was created in the upward movement of man and the labor system. Long ago, labor was a part of entertainment by verbal exchange and made its needs comfortable.

As time passed, it was transformed into an unbiased need to share one's soul, in sorrow or joy, or for no reason whatsoever, to drive out the soul, a need that is remembered day by day and used to be essentially moral and of psychological significance for the individual. Communication is such an essential issue of existence that stopping it will not be for people in any way for our animal ancestors; without the ability to communicate about an infant, one cannot probe about life and become a socially developed person. Depression brought about through loneliness also explains the best importance of interaction for humans. Using nothing for most is considered solitary imprisonment of criminals as one of the harshest punishments. In a scenario, he can transmit his intelligence to a person and sharpen his intelligence, but in the opposite case, he can also lose his reason.

A man or woman needs communication, some state of thought which he may be in pleasure or sorrow. But grief or suffering, which seek solace, sympathy, or just some distraction, are particularly challenging to bear alone. A person may additionally feel lonely and live among their family and are forced to lack association with pets.

Communication is not just a fundamental state of human existence; It is additionally a means of creating and developing social riding and moderation, which can also be realized through a backward person in the field of sport communication. Even when isolated, he understands his thoughts and actions from how others might react.

Historical development has drastically changed the means of influencing people's hearts and hearts. Speeches in the forum or senate, philosophers' conversations with their students, sermons in the church, singing songs, disputes between school students, legal professional, and public prosecutor speeches, professor's lectures, love letters, written proclamations, pamphlets, speeches of revolutionaries changed or replaced by mass media, radio or television, through massive adaptations of printed works has gone. Now streams of data are transmitted through the skill of qualitatively specialized channels across the planet, step by step integrating mankind through information. A splendid wealth of types of communication through the rich language of art, songs, poetry, music, paintings, testimonials, and novels is easy for humans. And how unspecified, infinitely rich, are forms of intimate communication. A psychological reaction or lack thereof is evident in facial expressions, postures, walking, gestures, changes in voice, hand actions, these are extremely used to express states of mind. In the whole machine of "body" language people search, especially those who experiment with such success, the inevitable situation is related to the eyes through which we produce each and all of it. Variety feels the glow of the human soul. Different intensities and perhaps even depths. What can anyone read in a face that does not have eyes?

Communication guarantees continuity in the development of culture. Every new technology starts its work of acquiring knowledge from the place where the previous generation left.

Thanks to the oral exchange of personal thoughts and aspirations, time is not denied help. They are equal in words, in images, they exist in legend and are handed down from century to century. Every man or woman bowed down to the ancient genealogical tree. The movement of thoughts in people's minds is like waves on the coast; behind them is the pressure of the entire ocean of world history. Books are the passports of the present for all the preceding cultures. In the treasure house of his original speech, era after era bears the record of the intense movements and events of perception. The whole impression of the intellectual lifestyle of man is written in words, in written characters, the invention of which human thinking solved the biggest and most difficult solutions to its problems. It incarnated, it registered the speech and, for this reason, acquired the ability to immortalize its ideas. "It is said that it cannot be erased with the help of an ax through a pen," is the folk saying. Writing is a wonderful and unbreakable fountain of information and knowledge, a fountain which in no way runs dry while it is in constant use. Communication occurs between specific people, eras, and also aspecific cultures.

Any idea of a communication problem inevitably leads to a query of mutual understanding. When it comes to understanding, one usually thinks of the understanding of real things, the cognition of the world around one. But what we are dealing with here is "communication understanding" how people hold each other using communication, how current technology is familiar to its predecessor, how humans of one culture understand other cultures. These are issues that have received little interest but are exceptionally important.

With the aid of events of telepathy, and so on, everyone is surprised by the combination sign. But some words as a whole are astonished through the "miracles" of communication, the language of words, gestures, imitations, and appreciation made through many symbols, mainly between the present and the past, and understanding between cultures. Mutual understanding at the level of continuous experience through communication seems to be mere trivialization given to the notion of one era or way of life through another. We understand all that we say and what other people, era, and culture say to us. And when understanding is not achieved, we often blame the language and speak that we are no longer able to find language consistently.

Attention was drawn long ago between understanding the objects and processes of the outside world and praising human tricks and words. In order for humans to understand and we have to keep in mind their objectives, the discrepancy between what they say and what they mean, we have to make allowance for the difficulties of locating proper motivation. There is a grand variety of individuals who stumble on mutual understanding. Each of us contains a complete world. And this world is our exact world. In any precise context of communication, a person usually exposes a component of himself. In a generalized way, we expand the understanding more using the experience of every other, with the help of our tendency to suit this tendency to positive and specific common needs that are unique in each person. The individualism of people's experience and frame of reference also makes mutual appreciation more difficult.

Sophist Gorgias once remarked that an object of expressing words in a system of thought disintegrates into a wide variety of factors of concept and consequently loses its integrity: the whole mutual understanding is therefore impossible. One regularly listens and reads, about the difficulties of interaction between adolescents and parents, between couples and between cultures, between healthy and sick people, especially those who are mentally ill. A foolish man or woman cannot fully express the views of the wise. From the content of what they have been informed, he absorbs only as fierce a form as he can understand. One must say that the diploma of mutual appreciation among humans depends to an amazing extent on their cultural level, the energy of their insight. The history of lifestyle exemplifies what mediums and instincts the power of the genius amplify through dealing with and resolving issues raised through the logic of life. Tasks of genius always embrace opportunities that have not appeared. And the diploma from which they are understood depends on the cultural degree of the reader, and the audience.

As the spirals of history rise, humanity continuously improves the mechanisms of mutual understanding, the content of interactions between eras and cultures. Each new era, in acquiring additional complete ideas, also acquires new eyes and looks into the magnificent works of the previous excesses and extras that are new, going deeper into their inner meaning. Many of Shakespeare's contemporaries saw him as one of the best, an attractive actor, and a little more. He no longer saw one of the supreme geniuses that humanity has produced, which has been proficient, printed by every new generation, century after century.

Wisdom alone cannot provide us with an understanding of a person, an era or a culture. There should also be a shared experience, the ability to empathize with different people, eras, and cultures. Where is the warranty that the present man is fully aware of the culture of the ancestors, their writings, paintings, sculpture? For example, a mere translation of ancient Indian writing into Russian cannot supply it. In order to fully understand them, one must make the way of life of human beings, in the socio-psychological context of each work, in life, every day, and the historical era in which it was written.

The personality of human members of the family depends to an incredible extent on this notion of every difference in the system of communication. If this is enough, the result is a vague relationship, whether that relationship is liked or disliked. Otherwise, the relationship becomes blurred.

Reasoning or proof is an essential aspect of understanding. Blank details cannot be understood by themselves. Another important aspect of mutual perception is the ability to listen. To do nothing, humans say that the art of listening is as important as the art of speaking.

It makes sense to understand the incredibly wide variety of different planes due to the reality that language and the total fabric of any speech context are linked to metaphors and threads of imagination. For the same reason, there is a phantom of regular understanding, as opposed to a factual notion of what is being said. However, despite all difficulties, the mutual oral exchange is built on a strong foundation of mutual understanding. Other than that, there should be no rational contact between people, and social lifestyles will be incomprehensible.

The team spirit of language and consciousness. If we favor the additional understanding of the interactions between people, eras, and cultures, then we should observe the nature of the media. Language concept is the best shape of expression, the basic ability to control behavior, to understand the truth, and the existence of oneself and culture. No person could collect cultural values without the gift of speech. Consciousness determines speech as its fabric fact in gestures, sounds, symbols, and similar structures. Speech can also express thoughts, feelings, and ambition in the method of mutual communication, due to the fact that words are physical and can, therefore, be perceived intelligently. Speech is the function of language in a unique state of communication. It is an attempt at verbal exchange and its recorded results. For example, Russian speech embraces all kinds of statements by specific individuals, and it is all written in that language. Language, on the other hand, has a distinctive vocabulary and grammar, expressed in proverbs and sentence patterns, which have traditionally been developed and are ubiquitous in character. But unique sentences, each spoken and written, are not of language, but of speech: they shape symbolic truths that constitute the existence of language.

Speech is the fabric expression of thought. The content of our mental world in speech is objective for others.

Speech fulfills quite a few interconnected tasks. It is both communicative and thoughtful. It can influence and regulate. The communicative feature is the main one. Since thoughts are non-physical in themselves, they cannot be considered with the help of experience organs. They cannot be seen, heard, tasted, or touched. The expression "people change ideas" is absurd if understood literally. There is no change in thoughts. The verbal exchange process is influenced by the ability of words to be shaped by a mutual fabric that seems to be an exchange of ideas. We no longer consider it using the ability of words; in the thinking of the person we are talking to, we generate corresponding thoughts.

With the ability of speech, a man or woman can internally manipulate things, their characteristics, and relationships, in their mind, touching or stopping them from seeing. Man has given this great boost due to language. It is simple to distinguish two aspects of the word: it's meaning and the structure of its existence. The first is a representation, an idea, an experience; the second is a symbol. An idiom is a harmony that has meaning and symbol. The phrase makes its meaning. A phrase does not represent an issue as a whole, but rather an element. A symbol is a textile object, process, a motion that exhibits a situation in the verbal exchange of representing something else and is used for receiving, storing, working, and receiving information. When we talk about the meaning of symbols, we have to think about data, things, their habitats, and relationships that are communicated to us through related symbols. It means it is a reflection of the objective truth expressed in the physical form of the symbol. Meaning, conceptual, erotic, and emotional components, unstable motivations, and requests, in short, occur throughout the field of consciousness.

The simple sign gadget is a common, everyday language. Non-linguistic signs and symptoms can also be classified as copy-signs (fossils of photographs, fingerprints, plants, animals, etc.), signs and symptoms as tremors (tremors as signs of disease, clouds as a sign of approaching rain), signs as signs (traffic lights, bells, applause, etc.), and signs as symbols. Consciousness is woven through many threads, forming a complex web of symbols, an entire and unique world.

Symbolism is an exact function of consciousness. It transcends all its boundaries and is expressed in the generalization which is the object symbol. For example, a flag is no longer a strip of cloth of a certain shade, although a piece of material with certain characteristics: color, shape, etc. What is a symbol? It is a certain object, action, process, phrase, or allegory, the meaning of which lies in the reality that they do something specific, including what they are, as it were, some other object or event. An image is an event that can also express a positive meaning now formally without delay. For example, justice is represented through the Devi doctrine. As a result, an image is no longer just a signal. In its external form, it is already an assumption, an image in which it is a symbol. A symbol has an expressive feature and, thanks to the embodiment of erotic concrete material, suggests something that is not in itself.

The use of specific symbols, and in particular the invention of artificial systems of formulas, is a blessing for science at large. In the scientific concept, the symbol system is characterized by the production of conceptual images. They contribute to the development of scientific perception towards an object and the creation of the world's true image. For example, the use of signs and symptoms or symbols from which formulas are made allows us to record the relationship between ideas in a concise form to increase the interaction internationally. Artificial sign systems, included in machines incorporating formal and code languages used in technology, are complementary to natural languages and exist only based on them.

Everything considered for humanity is named, imagined, or signified in some way. People have received an enduring need to know the names of things. Even when they do not accumulate any data from the identity of a certain person or object, they feel a positive pleasure in knowing what it is called or regularly show a deep curiosity concerning names, For example, getting the identity of a woman, or the name of a plant or distant star, even though it tells us very little.

Each word has a different shade of meaning, or even one kind of meaning, in a positive context due to the special individuality of cases and human situations. Its variations are as varied as the shades of color in peacock feathers.

The meaning of a word is "minimal knowledge," which refers only to the positive qualities of the object, rather than displaying its essence in all kinds. For example, when we are looking for that instrument of the word "water," we no longer reveal its physico-chemical form, we no longer explain the content of the given scientific concept (this work is the function of physics and chemistry); we indicate in basic terms that it is a liquid that is transparent. Many words can be used in a figurative sense. For example, the term "water" is sometimes used to refer to a lack of substance in a lecture, an article, a book, etc.

Although the feeling organs are once affected by speech, speech in itself, the fabric of its dress, is something that cannot be consciously considered. The speech focuses entirely on the object. Regarding reason, which considers things as opportunities in their admirable reality, it is neutral. We are confronted with a phrase or sentence, and a whole world of things and events arise in our head. A man or woman begins to follow phrases completely when he ceases to understand their meaning. Or he may also restore his thoughts on the phrase's fabric envelope, primarily for analysis purposes.

Accusing the role of only one car for thought exchange, it would be wrong to call the speech intellectually incorrect.

Speech performs an emotional, expressive, and regulatory-volatile function. Its emotional content material is pre-suppressed in a full range of rhythmic and stylistic instruments, in emotionally expressive terminology, in rhythm, stagnation, and exclamation in a variety of interrelationships. In relation to the whole complex of expressive movements, inclusive of gestures, facial expressions, and so on, as the ability to express speech.

The thought is usually an intellectual practice in any language. If a rational being from any other planet has been to visit the Earth and describe all the languages that exist nowadays, it should not fail to term their luminous similarity in logical structure, determined through the structure it happens—thinking of Integrated Earth Tools. If given thinking is expressed in English, Russian, or French, then the content of the three sentences remains the same, if not differentiated in linguistic form. The structure of language is formed under the decisive influence of target reality, through some unified needs of thought, and through the category shape of consciousness. But at the same time, these unified specific standards of thinking are catalyzed by thousands of different linguistic methods. Each nationwide language has its own individual structural and semantic features.

On this occasion, it is alleged that humans speak special languages that take up matters in an extraordinary way: this language determines the personality of perception. People classify things, their relations, and properties according to the current linguistic categories. Language, we are told, is not just about the content but also the size of the idea. Different people analyze the world in special ways. The structure of language completely determines the varieties of perception and behavior, and each language has its own philosophy.

Indeed, language has completely relative independence, its own internal logic. While the classes of recognition as complete have a simple personality (otherwise contact between different organizations would be impossible, and also translation would not be possible), the fundamental skill of expressing these classes is extremely diverse. There are currently more than 3,000 languages on the globe. This indicates the complexity and conflicting nature of the relationship between cognition and speech. In its structure, speech is not really a mirror reflection of the structure of the world of things, their homes, and relations; it is additionally a reflection of the mental world of the individual. It cannot consequently be equipped with thought, such as a cap on the head. Language affects cognition in the sense of its traditionally developed form, the unique nature of its semantic structures, and the notion of syntactic peculiarities with different colors. We are aware that the fashion of inquiry differs from the French in the German philosophical tradition, for example. Each style influenced different characteristics consisting of language, in two related peoples, and their nationwide cultures overall. On the other hand, any absoluteness of the effect of speech on cognition leads to the incorrect declaration that attention is directed not by the object, the target world, but by the way it is represented in language.

To do yoga, through the ability to speak, we talk to a person and inform him about our thoughts, feelings, and motives. We share the contents of our mental world. Consequently, speech has a certain intellectual content, which has to pass through language and come to phrases with its structure. Otherwise, if not rendered redundant, this material would count as an amorphous form that we would be unable to see as something with a particular quality. Linguistic size is no longer a prerequisite for expressing mere perception content; this is mainly a circumstance for the receipt of that material.

The relationship between recognition and speech is no longer truly co-existing and reciprocal, although a harmony in which meditation plays a decisive role. As a reflection of reality, consciousness molds "molds" and sets the legal guidelines for its existence in the form of speech. Consciousness is a continuously orally expressed reflection: if there is no language, there can be no consciousness. And any deaf-mute or blind deaf-mute who has had little training will reject this repeated principle: they have a kind of language of their own. And a complete lack of expertise can be maintained that these people barely assume based on visual images.

There is no case for the idea that awareness and speech live parallel, independent lives and only come together when a belief is implicit. They are two aspects of an important process: an individual thought using speech effort; he surprisingly performs speech activity. Think before you speak, popular knowledge says. If there is a thinking in our consciousness, it is always contained in a word, though it can no longer be the phrase that expresses an extraordinary idea. And conversely, if we notice a word, then an idea arises in our meditation with that word. When we are aroused using an idea, when a man or woman has a positive idea fully attracted to it, it "gets out of his head" inappropriate words.

Human thinking cannot ignore language barriers in search of this fact. Language is no longer the outward stage of thought, but the matter in which perception is settled. Naturally, the relation between consciousness and language should not be overseen, for example, through evaluating the idea of being a vessel's material language. This comparison cannot work, if only because the "linguistic vessel" is not empty in any way, no matter if its content is unfounded. Furthermore, the real intellectual content of the person does not exist outside the "language ware." Language is by no means exhausted through the outline of thought, and the concept is not separated from the language at any stage of its existence. Thoughts are no longer transformed into the language in such a way that their mental distinctiveness disappears.

Science records have made several attempts to identify thought and language to subtract from one to the other. These efforts are still being made today. For example, they are expressed in statements such as "statement is language" or "all philosophy is grammar." The concept of language as a noticeable summary shape, which includes a machine of familiar proverbs (universal grammar) forming linguistic sentences, suits very well with the general nature of thought, and this is something humans can do. This leads to the discovery of formal linguistic universality with a gradation of thought.

Consciousness signifies reality, although speech symbolizes facts and expresses a thought. Speaking is not yet thinking. It is a plateau, and it is valid only through regular life. If only one act of talking is to be thought of, as Feuerbach remarked, the greatest nonsense would be the greatest thinker. Possible knowledge, cognitive thinking;

Speaking skills communication. In a system of thinking, a person uses verbal material, and his or her thoughts are constructed, molded into speech structures. The work that is done to formulate ideas in speech is done extra or very little subconsciously.

When considering, a man or woman works on the cognitive content material and is conscious of it while the concept envelope may also be that awareness is managed in the backyard or on a fully time-honored plane. The thought was no longer to be imagined as a sort of "cloud suspended overhead" that opens and rains words. One cannot agree that the relationship between language and concept has come into fashion in such a way that, on the one hand, there are thoughts or ideas, that is, one that follows recognition and follows only by introspection. In contrast, separate hands are semantic structures, through the most important filters that ideas have to drop before they are embedded in sound. Speech serves not only to express but also to shape an idea. Thoughts are both forms in shape and speech.

The cohesion of consciousness and speech in the system of communication appears to be "self-evident." But is it possible for the idea that baring expressed in words exists? Processes of recognition that are no longer externally expressed take territory at the foundation of so-called internal speech, which is felt in the shape of internal interactions. The speech had to occur and mature as something external to emerge as something internal. When we accept silently, we regularly rehearse certain thoughts in our minds unconsciously. Internal speech is voiceless. It is diverse and concise of external speech. Meditation, which takes place in the structure of internal speech, is a type of dialogue with oneself regularly. This type of speech is purely an imaginative, communicative function, and its primary function is an instrument for formulating and forming ideas. Internal speech is no different from external speech, but by its characterization entirely through its structure. Since the inner voice is targeted at itself, it leaves out the whole thing that can be understood.

Is perception possible except in speech? We emphasized above that there used to be insoluble solidarity between awareness and speech, which is real as a general rule. If it has been viable to express everything in words, why should there be expressive movements, plastic art, painting, and music? And how do things stand in relation to scientific theoretical thinking? As Einstein informed us, the positive moments in the mechanism of his symbolic activity everyday, as noted and written, played no decisive role. He used to be able to assume more or less vivid pictures of physical reality: the sea in action emblematic of electromagnetic waves that cannot be experienced visually, the physical force moving in the same manner as the work of muscles, and so on.

And how does the work of thought take the field when one is swept away by the light of fact on the "wings of intuition" and not by the "rope ladder" ability of logic?

This is not just because the conceptual thinking method is constantly intertwined with an imagination that seeks no verbal form. Thinking in pictures can also be deeply conceptual because pictures can also enrich the status of symbols rich with conceptual content. Generally, no one has yet been able to prove from the data that the concept takes place only on the basis of natural language ability. It is only stated, but the journey tells us otherwise. However, the thinking in Pix takes the field as an exception or in the structure of elements woven into the content of the routine cohesive activity and does not dismiss the popular theory of the unity of consciousness and speech.

We know that the concept's possibilities are definitely in line with the possibilities of the given language: a poor vocabulary is a positive sign of mental poverty. This is quite natural. A person can work with the fully accumulated understanding that is enshrined in the semantic word of the language. The primitive man, who used to be disgusting, used only a few dozen words, while these days the average character has an active vocabulary of 3,000 to 5,000 words, and the lead author uses over 10,000. Nevertheless, poverty in the intellectual sphere does not detract from bad terminology. On the contrary, poor vocabulary is the result of superficial thinking due to a lack of culture, social travel, and social relationships.

One of the concrete arguments for the principle of harmony of thinking and words is found in scientific facts, which informs us that intellectual disorders affect speech.

In simple cognition, the method of verbal exchange seems to be very simple, some things which can be taken as the number of courses. But the expression of focus in phrases is regularly an extremely complex problem, and not every speech system quality of ideas is possible now. We often feel that what we have said does not adequately express what we are thinking. We reject one phrase because we no longer completely substitute our thoughts for the other. The content of thinking controls his oral expression skills. A man or woman cannot once remember a phrase or name, even if it is "at the tip of my tongue." But whatever is properly conceptualized is clearly expressed. A pleasant assumption is devalued using a bad assumption. There are two types of nonsense: one comes from a lack of concept and feels hidden through words, the other, from an overabundance of thought, and important words are not remembered to make them clear.

The realization of the process of thinking in forms of language involves the discomfort of mental creativity and the pain of finding it a uniquely adequate skill. Sometimes an idea that is on focus can also be for a time, as Mayakovsky said, "authorship sucks." Thought must overcome positive external material, which is resistant to thought now and again.

We know that language breaks old forms of thought. To recognize the world of these days, we use words created through the world of tomorrow. In addition, language affects consciousness in the sense that it forms a coercive, "tyranny" over thought, directing it along positive linguistic channels, constantly changing, in my opinion, its boundless thoughts and, in general, linguistic, emotionally colored ideas with nuances and patterns, thus putting a kind of embryo of universality on the idea. Sometimes it throws away the notion of cliché and hackneyed phrases for mercy.

The more unusual our experiences are, the more difficult it is to categorize them with the help of socially developed, planned symbolic means. Platitudes are expressed more readily; they are like a favorite float of metal, which comes free in a cliche drawn from the language. Thoughts, feelings, and speech all have a male or female character.

When we talk about the language of Pushkin, Shakespeare, or Gogol, we generally think of linguistic competence and the unique approaches used by these authors. A person can also judge a character with the nature of his speech with the help of multiple gestures, the way he talks. There is a closed relationship between the way to question and the way to express ideas. If, for example, we study the creative techniques of any writer, we quickly conclude that the old and painstaking work on the shape to which the concept is expanded furthermore makes the idea fuller and sharper. It also works on doing. A basic rule for almost any creator is to rewrite, revise, insert, and usually change his or her manuscripts. Dostoyevsky claims that a writer's greatest ability is his ability to transcend it.

Speech has a powerful ability to influence human psychology. And this feature is one of its oldest. A well-folded phrase can sometimes leave soldiers in flight and snatch victory from defeat. A word can be a medicine that relieves human suffering or a poison that induces euphoric pain. So there is a ton of energy to influence in language. We all believe in the power of words. They can force a person to cry or laugh. Words can kill a person and comfort him in his grief. In historical times, when everything prevailed with a belief in the magic of words, and even today, words are recognized to have a kind of mystical powerful influence and therefore use through psychiatrists skilled in the restoration of their patients is done.

The purpose of oral communication is no longer a complete understanding and settlement, but moreover, a desire to advocate for someone else, explain it, teach, influence the person and guide his/her actions. So-called volatile relations exist among humans, expressed as commands, instructions, prohibitions, permission, obedience, disobedience, and the like.

Chapter 5.

Foster the willingness to accept a painful experience

"Pain is unavoidable, conflict is optional"

Survivors and adolescents from extreme and complex trauma can be difficult to care for, and it can be difficult to find peace amidst the chaos. When we enter survival mode, we may lose deadlock, and circumstances may soon become out of emotional control.

What has been done, and what is the critical importance of it?

Everyone experiences pain. Pain is no longer an optional segment of being alive. pain, however, is often temporary. This often creates feelings of sadness, despair, loss, and frustration. On the other hand, grief is caused by a refusal to be given the truth because it is in a deep and extended state of distress, helplessness, anger, hatred, revenge, or despair.

The top information is that conflict is optional.

Radical acceptance is about reality, not liking it or hostility against it. Radical acceptance has nothing to do with passivity or giving up. On the contrary, it is in pursuing its power rather than being trapped in misery.

When you fundamentally achieve a painful reality, your thoughts, feelings, and attitudes must shift to make room for what has been happening (or has been a Beefel in the past).

Pain + accept = pain two give two pain + non acceptance = pain

When an event is painful, it is to try to push it away, fight on its behalf, attempt to manage it, or numb it using unhealthy copulatory behaviors such as alcohol, drugs, self-harm, blame, aggravation. Avoiding, shopping ... These coping strategies are deceiving, because as long as they all give very intense, immediate, and predictable (temporary) relief, then dismissing reality only intensifies our emotional reactions. As a result, the time spent on unproductive thoughts and behaviors is spent.

Sadly, these spot relief strategies can no longer change the reality you are trying to escape/deny. It will most likely make the state of affairs extra painful and challenging (without any submission to permanent rest or resolve). It results in a waste of time and little power for no return!

Components of technical acceptance (what it is)

Accepting things accurately and just the way they are, accepting what we cannot control.
Being factual about the situation (adding additional incorrect or exaggerated meanings).
Accepting FACTS as they are, including our own work
Tolerate feelings/thoughts that come from pain and know how to live with things that are no longer far from the present.

Components of medical approval (what it is not)

Radical acceptance does not agree with all that has happened
Radical approval is no longer approval
It is no longer condemning the conduct or tricks of others
It is no longer accepting your needs; acceptance is no longer the same agreement
It is no longer ignoring or denying a situation; it is no longer surrender

Choose four sessions

Solve the trouble (if you can). If you can't trade it, you can:
Change how you feel about trouble (making lemonade), or accept as it is, or being sad
This is a priority and my own choice to make it: both measures redefine it, deliver it, or continue to grieve.

Why should I believe

Rejecting the truth no longer trades reality
Pain is inevitable, and it is a part of life
Refusal to accept reality leads to sorrow: extended and widespread grief, bitterness, anger, shame, thoughts of revenge, and anger acceptance, while regularly coupled with sadness, creates an area for deep peace.
Because life is worth living even when it involves traumatic events
Manufacturers are tentatively accepted with profound effects.

No skills:
Mindfulness takes a look at the potential and describes what has happened completely and non-judgmental to enable time to reduce thoughts.

Disturb tolerance abilities to manage uncomfortable and painful emotions that occur without worsening the state of affairs.

Emotion regulation is the ability to deal with strict and uncertain thoughts, and the emotion remains away from the mind reacting fully.

Interpersonal efficacy competencies include interacting with others in productive, truthful, and respectful ways, starting therapy.

Trying not to "get off the hook" to another man or woman, holding anger can make it seem like you're punishing the offending person. As long as you remain angry, whatever they do to harm/annoy you, they are not overcome. Your anger serves as a memorial to what happened.

Confident that I agree with what happened while acknowledging the potential. Radical acceptance is accepting the match as true, real, and has meaning for you.

You need to be irritated to defend yourself. Radical acceptance can seem risky, taking away your anger, withdrawal, and resentment, which have served to save you from pain. Non-acceptance exercises a resounding deal of power and center on the state of matters that have already occurred and cannot be changed.

Radical acceptance has nothing to do with the other person. It is entirely about reducing their pain and removing extended suffering.

Radical acceptance helps you move forward and still holds different people responsible and fully accountable for their behavior.

Practical accessories, step-by-step

Observe if you are at war or questioning the fact ("It should not be that way" "It is not happening" "It needs to be stopped").
Remind yourself that this is the state of affairs and cannot be changed ("This is what happened" describes the situation factually and accurately).
Remind yourself that there are reasons for what happened. There was some history in this match or situation. Notice how, given the record and the causal factors, this truth had to be exactly the same as it did. "I hope this can happen," given what I know.
Practice accepting self-talk, relaxation, emotion rule skills, prayer, or mindfulness to soften the pain.
Allow any frustration or sorrow to arise and be present apart from actually facing it or making it worse.
Accept that life can be desirable even when there is pain.

Chapter 6.

Discover the very important things in your life

We live in a cramped world using static deflections. There are endless activities going on around us at all instances that can grab our attention, and often those who win loudly shout. A simple way to enhance your existence is to explore your person.

Most humans live their lives by focusing on what they have to do. Countless tasks continue to grow, and we wonder why we do not feel in any way as we are moving forward. It seems that we are just trying to run on the treadmill, and every project done has been quickly replaced with the help of new people.

Life is much less difficult when we stop asking ourselves why we do things.

What is the purpose of your life?
Now, this is no longer some largely esoteric question that you will ponder for a lifetime. These are just a few things that you set for your own and can trade at any time. What is the list of matters that are most important to you in your life?

I think it helps to write the things that you are actually doing. It may be spending time with friends and family, an exact hobby, your job, or traveling. The solutions are going to be exclusive to every single man or woman reading.

What's important ... What is required of you that is very difficult to improve your existence if you are not aware of what honesty means to you personally?

Refining your list of priorities
Once you have your initial list, you want to see what is required of you. You do it by asking yourself why you love it and why it is necessary to keep it in your life. I was doing this exercise in the final three years, and I came to know that as I continued, my list became extra and sophisticated.

You also want to show how well you spend time on these things that you love. Do you get to spend a lot of time with each one, or have they been pushed again behind other duties in your daily life?

Discover what's really important
If you have made some other list of all the things in your life that you are actually doing, you will likely discover a bunch of things you just don't love. Some people may hate playing football, while others will like it. Everyone is unique, and it's about finding out what you like to do... now it's not what you're experiencing.

It is easy to fill our lives with things that certainly do not matter to us. The trick is to spend less time doing unimportant things and more time doing things that we love. This is no longer an ideal technique for leaving everything out for now, but as you draw attention from this point on, you will be amazed at the results.

Always ask why

Whenever a new item arrives on your plate, you should ask yourself why it is important for your child's faculty to be part of that new committee or be more positive to spend that time with your kid instead? Does it matter if you miss your health club time at lunch because your boss wants you to work extra time, or is your fitness extra necessary for you?

Our world is full of expectations that the help of others satisfies. We wish to be gorgeous human beings who complete a round of exceptional cases each day and every time; to be of high quality at work and home. When people ask how you are, and you do not give a "busy" answer with the silliness that appears in your eyes, you notice that people often look at you uncommonly.

However, once you start analyzing your lifestyle, why, you start asking yourself, "What's important to me?" You want to break away from society's expectations and instead start paying attention and refine your expectations. When you do this, you will see the world from an utterly exclusive point of view and will be free to improve your life.

Many businesses and enterprises offer organizational capability training, whether it is a workshop, enterprise presentation, line coaching course, or an all-out conference. Participating in these opportunities is ideal for gaining knowledge of organizational skills. Then, of course, you can set your personal goals.

For most people, organizational skills do not come naturally. However, fortunately, like any other skill, they are learnable. Once you gain an understanding of a skill, the more you practice it, the more you will achieve it.

If you are new to all of this, then starting small is your quality bet. Set yourself a goal, choose a factor on which you want to improve and repeat it several times until it becomes a habit. Once you are convinced to maintain this habit, you can add to your goal or grow on it.

Starting small and taking steps as you progress is an appropriate path of action, as it can ensure that you honestly achieve what you have accomplished. If you dive straight into the deep end, you have a chance to become more overwhelmed than before and may even fail to meet expectations.

Surrounding yourself with people who have accurate behavior is another way of studying organizational skills. Being the leader, manager, or head of a commercial enterprise of a great organized group can greatly affect your own movements and behavior.

Ten organizational skills training techniques
If you have recently felt yourself-overwhelmed and upset at work, you should probably try one of the following organizational competencies coaching techniques. They should help you regain control over your tasks, regain focus, and reduce stress levels.

1. Make a list
If you are feeling affected by tasks, then developing a to-do list is great for taking control of the things you need to do.

By writing down your duties in order of importance (make sure you have prioritized your list!), you will get an idea of what needs to be done.

2. Do not trust your memory
Even if you have a supernatural memory, it is right thinking to write everything down completely.

From the challenge deadline to customer details, to product prices, the following can serve as a reminder, so that you don't forget about essential matters when you're feeling overwhelmed.

And since most of us have smartphones, you can by no means go wrong with writing things down in the device.

3. Schedule
A large part of being equipped is to know how to plan, and expert planning involves too much scheduling.

Scheduling is one step ahead of the to-do list. On the whole, you do not have the cases you need to file, but you have a timetable to complete them. This helps you develop your time administration skills, as you anticipate coordinating tasks and things so that cut-off dates are completed, and the whole thing is executed on time.

4. Learn to delegate
Learning to delegate tasks is a valuable ability that will help keep you organized. This will not only lighten your workload but also speed up your planning and prioritizing ability as you have to study which tasks you will do and which tasks are okay to be given to someone else.

5. Avoid multitasking
Although an attempt to undertake more than one venture at a time may be thought of as brilliant in practice, it is the

exact opposite. Multitasking is definitely recognized to reduce your productivity because it reduces your focal point and interest, and things become more challenging and take longer to complete.

6. Reduce Interruptions

It is impossible to manage every aspect of your environment, but it does not hurt to try. Reducing interruptions while you are at work gives you a better chance to eliminate them as accurately and successfully as possible.

Investing in noise-canceling headphones or putting social media blocks on your desktop are examples of ways you can reduce distraction.

7. Reduce clutter

An excellent organizational capability coaching technique is to have a tool for your documents. Whether it is at home or work, we all submit files that we do not currently need, but are afraid to throw in case we want it in the future.

Having a well-equipped device allows you to locate the original archives you want at any time. This keeps them safe, which reduces the risk of losing something. This submission system applies to both actual paperwork and digital documents.

8. Organize Your Work Area

Where we work greatly affects how we work. If you have a cluttered and messed up workspace, your chances of being unorganized may be high.

Having an organized workspace ensures that you are capable of doing your most productive work. You have not wasted time looking for matters that have been done wrongly, and working in a clutter-free environment can be pleasing to your mind.

9. Discard unwanted items
Disorganization is considered a cause of stress and anxiety. [2] If you are already feeling overwhelmed, then the sight of garbage can increase that feeling.

Get rid of the things you no longer want to clean your environment and, hopefully, your thoughts too.

10. Keep fit regularly
While working, it can be convenient for your desk. You are targeted at work, and therefore maintaining the entire lot at your desk is probably a low priority. But for this, one must be conscious. A simple tidiness may cause something wrong with your desk.

Whether it is a quick smoothie every day, or deep cleaning every month, being conscious of tying and fitting it in your activities will help you stay organized and reduce stress.

Chapter 7.

Commit to living a meaningful and vital life.

Parents regularly say: I only choose my children to be happy. 'It's unusual to hear: I'm just in favor of making our children's lives worthwhile,' but most of us want that for ourselves. We worry in vain. We fret about the nihilism of this or that factor in our culture. When we lose our sense of meaning, we become depressed. What do we name this issue, and why do we need it so badly?

Let us start with the ultimate question. Certainly, happiness and meaningfulness overlap regularly. Perhaps some degree of meaning as a condition for happiness is a reasonable but insufficient condition. If this has actually happened, then humans will consider other humans only as a means to step in the path of happiness. But then, is there any reason he wants for his own good? And if there is not, then why would man ever choose a life that is extra meaningful than being happy, as they do from time to time?

The difference between meaningful and happiness used to be the focal point of an investigation that I posted in this August's *Journal of Positive Psychology* with my fellow social psychologists Kathleen Vohs, Jennifer Acker, and Emily Gurbinsky. We surveyed about four hundred American citizens, ranging in age from eighteen to sixty-eight years. The survey raised questions about how people thought their lives had been blissful and the extent to which they thought they were worthwhile. We no longer defined happiness or meaning, so our subjects responded to their own use of those words in their own sense. By asking a wide variety of questions, we were able to see which elements went with pleasure and meaningfulness.

Subscribe to our newsletter
As you might expect, the two states overlapped substantially. In an almost meaningful sense, half of the version was explained through pleasure, and vice versa. Nevertheless, using statistical controls, we were able to manipulate the two, separating the 'pure' results of each, which were no longer entirely based on the other. We limited our search to factors that had opposite results on happiness and meaning, or at least, factors that had a high-quality relationship with one and did not even indicate an amazing correlation with the other. Using this method, we discovered five sets of the most significant differences between happiness and expressiveness, with five areas being company-specific versions of people with precise lifestyles.

First, you had to choose and get whatever you needed. Not surprisingly, the pleasure of desires was once a reliable source of happiness. But there was nothing in it - perhaps much less than nothing - to add to the sense of meaning. People are happy to such an extent that they find their lives as the most difficult choice. Happy people say they have enough cash to buy the things they want and the things they need. Good health is an element that contributes to happiness but is no longer meaningful. Healthy humans are happier than sick people, but there is no lack of meaning in the lives of people with poor health. The more and more people feel good - a feeling that can arise from achieving a wish or desire - the happier they are. The more often they feel bad, the happier they are. But the frequency of sublime and terrible feelings becomes irrelevant to meaning, which can flourish even under very forbidden circumstances.

Social lifestyles used to be the location of our 0.33 group of differences. As you would probably expect, connections to other people became essential for both meaning and happiness. Being alone in the world is associated with low levels of happiness and meaningfulness. Nevertheless, it was the unique personality of a social connection that decided which state they helped to describe. Meaningfulness comes from contributing to different people, while happiness comes from what they add to you. This is in contrast to conventional wisdom: It is widely believed that helping various humans makes you happy. Well, the extent to which this occurs depends entirely on the overlap between meaning and happiness. Helping others was a tremendous contribution to the meaningful fairness of happiness, but there was no indication that it promotes happiness independently. If anything, its effect was in the opposite direction: as we propose for the true meaning to enhance it, helping others can completely detract from one's own happiness.

We saw echoes of this incident when we asked our subjects how they took the full time to care for the children. For non-parents, childcare did not contribute anything to happiness or meaningfulness. Taking care of young people for most people is obviously neither very good nor very unpleasant, and it is not important either. For parents, at different times, caring for the youth was a great source of meaning, it appeared next to the point of happiness, presumably because the youth are happy from time to time and when it is stressful and annoying, it creates balance.

In our survey, humans were valued as 'givers' or as 'takers.' Giving yourself as a person strongly presupposes more meaningfulness and much less happiness. The taker's results have been weak, perhaps because people are reluctant to accept that they are the taker. Nevertheless, it was once clear that a taker (or at least, considering oneself as one) promoted happiness but diminished its meaning.

The depth of social relationships may also differ from how social existence contributes to happiness and meaning. Spending time with friends was once more associated with happiness, but it was inappropriate for meaning. Having a few beers with or participating in fine lunch conversations with friends will likely be a source of enjoyment, but overall, it is not very important for a meaningful life. By comparison, spending more and more time with loved ones was once associated with a higher meaning and was irrelevant to happiness. The difference, of course, lies in the depth of the relationship. Time with friends is often committed to simple pleasures, which prevent tones at stake, fostering top emotions while doing something to expand meaning. If your friends are grumpy or tedious, you can just move on. Time with nurtured people is not so equally enjoyable. Sometimes one has to pay bills, deal with diseases or repairs, and do various unsatisfactory work. And yes, nurturing can also be difficult, in which case you usually have to work on the relationship and pull it out. It is probably no twist of fate that arguing itself meant to relate more meaning and much less happiness.

The fourth category of differences had to do with conflicts, problems, tensions, and the likes. In general, these tended to decrease in pleasure and with greater meaning. We asked how many negative and positive events humans have experienced these days—having a lot of good cases made both means and happiness beneficial. No shock there. But bad things were a story of sorts. Highly meaningful lives come on a lot of negative events, which limits happiness. The events of stress and bad existence were two powerful shocks to happiness, regardless of full-size fine engagement with remarkable life. We start feeling blissful, but now there will not be a significant existence. Reflecting tensions, problems, anxiety, arguing, challenges, and conflicts - all these are little or absent from the lives of happy people, but they seem to be part and parcel of an important life. The transition to retirement illustrates this difference: With the end of work demands and stress, happiness increases, but there is a meaningful decline.

Do humans go out in search of stress to add that tool to their lives? It is more likely that they seek meaning through the pursuit of difficult and uncertain tasks. Tries to meet matters in the world: it brings each fluctuation, so the net acquisition for happiness will probably be small, although both contribute in a meaningful way. To use an example closer to home, conducting research greatly adds to the sense of a significant existence (what could be more important than working to make the savings of human knowledge larger?). But projects rarely happen as planned, and many failures and frustration along the way can suck some happiness out of the process.

The last category of variations had to do with self and non-public identity. Activities that make themselves distinctive are an important source of that means but are usually next to the point of happiness. Of the 37 items on our list that asked people to charge that some effort (such as working, exercising, or meditating) was self-expression or reflection, 25 had a good correlation with a meaningful existence used to produce and nobody was once negative. Only two of the 37 gadgets (socializing, and partying without alcohol) were positively associated with happiness, and some also had fairly terrible relationships. The worst was once a concern: if you consider yourself as a barrier, it seems to be substantially inferior.

If happiness is what you want, then it seems that meaningfulness is about doing things that make themselves clear. Even just caring about personal identity and self-definition problems, was once associated with extra meaning, even if it was irrelevant, if no longer outright harmful, to happiness. This may sound almost contradictory: happiness is selfish, in the sense that what you choose is about you and other human beings doing things that benefit you but are more tied to happiness. Expressing oneself, defining oneself, building a top-notch recognition and joy rather than happiness from various self-oriented activities means more and more.

Does all this tell us something about the meaning of life? 'Yes' answer depends on some logical assumption, at least not now the concept that man will tell the truth about whether his life is meaningful or not. Another assumption is that we are also capable of giving real answers. Can we know if our life is meaningful? Shouldn't we be in a position to say what it means? Remember that my colleagues and I did not give our address to the definition of meaning to the respondents, and we did not ask them to define themselves. We asked them to rate their degree of agreement with statements such as: 'In general, I think of my existence as being meaningful.'.

First of all, what is life? An answer to the title of the novel *A Vitro Phenomena* (2013) in the wake of Anthony Marra about Chechnya after recent wars gives resources. A man is stranded in his union for doing nothing and starts analyzing his sister's Soviet-era clinical dictionary. This affects him little as useful or even understandable statistics - moreover to the definition of life, which he sees in red: 'Life: a constellation of important events - organization, irritability, movement, development, reproduction, adaptation.' In one sense, 'life' means. I must add that we now recognize that this is a specific type of physical process: no longer atoms or chemical substances themselves, although the organized dance they perform. The chemicals in a physique are very much equal, especially from the moment of loss of life to the first after the second. Death does not change this or that substance: the entire dynamic state of the body changes. However, existence is a purely physical reality.

That meaning of 'Earth' is more complex. Words and sentences have meaning, as is life. Is it the same type of issue in both cases? In a sense, the meaning of 'life' should be a simple dictionary definition, as I gave in the previous paragraph. But when asked about the meaning of life, the man no longer chooses, anyone greater than this will assist anyone who was suffering from an identity crisis to study the name on his driver's license at some point. An important difference between linguistic meaning and human existence is that 2d implies a cost judgment, or a cluster of them, which implies a certain form of emotion. Your arithmetic homework is loaded, meaning that it is entirely a network of concepts, in other words. But in most instances, there is no great emotion to do so, and so people do not tend to consider that feeling in which we are interested. (In fact, some people are doing mathematics, or are curious about it, although those responses rarely appear conducive to viewing anxiety as a means of supply in life).

Linguistic meaning in various types of non-physical relationships. The two cases can be physically connected, for example, when they are held together, or when one of them draws gravity or magnetic over the other. But they can also be added symbolically. The unity between a flag and the United States represents that there is no physical relationship, molecule to molecule. If the country and the flag are on opposite sides of the planet, it remains the same, making direct physical connection impossible.

The human mind has evolved to use meaning to recognize things. It is part of the human way of being social: we talk about what we do and experience. Most of what we recognize was from others, not from direct experience. The study of our very remaining language depends on cooperation with others, ethical and legal policies, and so on. Language is the device with which people manipulate meaning. Anthropologists love to find exceptions to any rule, but they have failed to discover any subculture that occurs with language at some distance. It is a human universal. But there is a big difference to make here.

Although language as a total is universal, unique languages have been invented: they change through culture. Its meaning is universal, yet, we do not invent it. It is discovered. Consider mathematics homework: Symbols are arbitrary human inventions, although the idea expressed by five x eight = forty-three is inherently wrong and is no longer something that humans can create or change.

Professor of psychology at the University of California, Santa Barbara, neuroscientist Michael Gazzaniga, referred to the term 'left-brain interpreter' as an aspect of intelligence that appears to be almost entirely devoted verbally to accomplishing it. The left-brain interpreter's account is no longer always right, as Gazzaniga has demonstrated. People increasingly formulate explanations for something or experience that shape the narrative, to shape their story. His errors led Gazzaniga to question whether the process had any value, but perhaps his disappointment was colored through the scientist's natural belief that the purpose of thinking is to exclude the truth (this, after all, is what is believed). Conversely, I believe that there is a greater class of thinking due to helping to talk to different people. The mind makes mistakes, but when we discuss them, different humans can recognize mistakes and correct them. By and large, mankind strategizes reality as an alternative, rather than by matters of thinking through alone, by collectively deliberating and arguing.

Many authors, especially those with meditation and Zen travel, have an overview of how human thought manifests throughout the day. When you try to meditate, your mind is heavily influenced by thoughts, once known as 'internal monologue'. Why does it do this? William James, the author of *The Principles of Psychology* (1890), states that it is improper to inquire, but in fact, many questions seem inappropriate. However, putting our ideas into phrases is a mandatory directive to convey these ideas to other people. It is important to talk: it is how a human creature connects to and participates in its group - and it is how we solve the eternal biological issues of living and reproduction.

Man developed the mind that talking nonsense all day is because it is how we survive. To talk, a human being has to do what he does and write it in words. A deer can walk down the hill and get a drink, as can a person, though only one person thinks the words 'I'm going down to drink.' In fact, humans will no longer presumably eclipse those people. However, when you say the words out loud to them, others may come for a day trip - or perhaps offer a warning not to go after all, because someone saw a bear on the water's edge. By talking, humans share information and connect with others, which is about all of us as a species.

Studies were done on teenagers' with the idea that the human mind is naturally programmed to put matters into words. The children loudly say the name of everything and are trying to name all kinds of things, such as shirts, animals, and even their bowel movements. (For a time, our young daughter was naming hers after a series of relatives, with no animosity or disrespect, though we did not inspire her to name it). Bids of this type would be a problem at once. Solutions are not useful. The familiar uses practical thinking, although it helps to translate the physical phenomena of one's existence into speech so that they can be shared and mentioned with others. The human mind evolved to engage in collective discourse and social narrative. Our tireless efforts to make sense of things begin with small gadgets and events. Very slowly, we work towards larger built-in structures. In one sense, we climb its ladder, which means - from single words and concepts to easy mixtures (sentences), and then grand narrative, comprehensive philosophy of cosmic theory.

Democracy exposes how we use meaning. It no longer exists in nature. Each year, countless human organizations make behavioral choices, but none have yet seen one in any species. Was democracy invented or discovered? It presumably emerged independently in many particular places, although the underlying similarities suggest that the idea was once there, and was designed to be found. The exact practice (e.g., how the vote is taken) is invented to implement it. All the same, it seems as if the thinking of democracy was just ready for a man to stumble upon it and put it to use.

Thinking about the meaning of lifestyle implies that there is a long way up the ladder. In order to understand the meaning of some newly encountered items, the human may ask why it was made, how it is received, or what is its usage. When they come to the question of the meaning of life, comparable questions arise: Why or for what reason was life created? How did this existence get here? What is the proper or good way to use it? It is natural to anticipate the answers to these questions. A child learns what an orange is: it comes from the shop and, before that, from a tree. It is suitable for eating, which you do through the outer (peel), to get the soft, candy inside. It is natural to count life that can be understood equally. Just find out (or research from others) what it is about and what to do with it. Go to school, work, marry, have children? Sure. Also, there is a top motive to get it all straight. If you had an orange and failed to recognize it, you probably wouldn't get the benefit of having it. In the same way, if your lifestyle had a purpose and you don't know it, you can stop losing it. How sad to leave the meaning of life, if there is one.

We begin to see which means of life thinking holds two specific things together. Life is a physical and chemical process. Meaning is a non-physical relationship, something that exists in a network of symbols and references. It climbs an extraordinary distance to connect through home and time because it's not purely physical. Remember our findings of the unique time frame of happiness and meaning. Happiness can be closed to physical reality, as it currently is right here. In an important sense, animals can be comfortable without much in the way of meaning all the way through. Meaning, in contrast, connects past, present, and future in ways beyond physical connection. When the cutting-edge Jews had a Passover day, or when Christians symbolically feasted through swallowing blood and eating the flesh of their deity, their footsteps were guided by symbolic connections to events in the distant past (Actually, events whose very reality is disputed). The link from the past to the present is not one of physical form, in the way a line of dominoes falls, although alternatively, it is an intellectual connection that leaps over the centuries.

Life questions which are more driven by mere passive curiosity or anxiety about lack. Artha is an effective tool in human life. It helps to talk more about existence as an ongoing change system to identify what that tool is being used for. A survival factor can usually be in flux, although lifestyle may not be at peace with endless changes. In their quest to establish a harmonious relationship with the environment, survival is a yearning for sustainability. They want to explain how to get food, water, safe haven, and so on. They create or find places where they can rest and stay safe. They will likely hold the same homes for years. Life is, in other words, an attempt to end a dull routine or change-of-way, which leads to the end of death. If the exchange should stop altogether, it was mainly on some of the best points: once the subject of an intense story of Faust's wager with the devil. Faust misunderstood his soul because he did not want a brilliant second to be closed forever. Such desires are meaningless, and they can't stop changing until life ends. But residency things work hard to establish some sustainability diplomas, reducing the chaos of regular trade to a somewhat stable reputation.

Conversely, the meaning is mostly fixed. The language is completely practical because it has the same meaning for everyone and the same meaning for the next day. (Languages change, although slowly and indeed reluctantly, relative stability is integral to their function). This means that it consequently presents itself as an essential tool by which humans balance their world. By recognizing the continuous rotation of the season, people can sketch for future years. By establishing permanent property rights, we can further expand farms to grow food.

Naturally, humans work with others to apply their meanings. Language has to be shared, and there are no longer real languages for private languages. By speaking and working together, we create a predictable, reliable, honest world where you can take a bus or plane to go somewhere, believing that food can be purchased after Tuesday and that you won't have to sleep outside.

Marriage is a good example of how meaning diminishes the world and will increase stability. Most animals have intercourse, and some do so for a long time or even for life, although only humans marry. My co-workers who are in close relationships will tell you that relationships develop and change after many years of marriage. However, the truth of marriage is constant. You are either married or not, and that does not fluctuate from day to day, even if your feelings and your move towards a spouse changes your feelings. Marriage smoothens these bumps and helps to stabilize the relationship. One reason for this is that humans prefer to live together if they are married. Tracking all your feelings towards your romantic partner over time will be difficult, complicated, and potentially incomplete. But knowing that you have made the transition to not being married is easy, as it happened on a specific occasion that was officially recorded. Meaning is more stable than emotion, and therefore shows that the use of living matters as the stage of their endless quest to achieve stability.

The Austrian psychoanalytic thinker Viktor Frankl, the author of *Man's Search for Earth* (1946), attempted to replace Freud's idea to connect Freud's specific desire for other drives to economies. He emphasized a sense of purpose, which is certainly a factor but possibly no longer the whole story. I have my own efforts to find out how people discover that the meaning of lifestyles has finally settled into a list of four' meaningful needs, and in later years, the list has been quite good.

The key point of this list is that you will find the lifestyle meaningful to the extent that you have something that addresses each of these four needs. Conversely, humans who fail to satisfy one or more of these desires probably find life to be less than worthwhile. Changes in relation to any of these desires should also affect how important men or women find their lives.

The first is for the purpose, indeed. Frankel was once right: in the absence of purpose, existence lacks meaning. A motive is a future state that shapes the present, thus adding extraordinary examples to a story. We can divide the objectives into two broad categories. One can strive towards a unique intention (to win a championship, to become a vice president or to raise healthy children) or a state of fulfillment (happiness, spiritual salvation, financial security, knowledge).

Your Free Gift

In order to thank you for your purchase, we're offering a free gift exclusively for the readers of **Acceptance and Commitment Therapy: The Action-Oriented Psychotherapeutic Approach to Reducing Stress, Anxiety, Panic Attacks, & Depression**.
An introductory chapter of my book: **Dialectical Behavior Therapy**.
By downloading this chapter, you will understand better some concepts on the topic of DBT.
Click here to access your free gift

Did you know that can download the audiobook of this book for free?
Click here for Audible US
Click here for Audible UK
Click here for Audible FR
Click here for Audible DE

Don't forget to leave a review on this book. Is simple! Just click on this LINK and you will be directed to the right page. It's very important for me. I will appreciate if you do.

Page intentionally left blank

Dialectical Behavior Therapy

Master Your Emotional Intelligence with DBT, Control Borderline Personality Disorder and Overcome Depression, Anger and Panic Attacks

Leona S. Murray

Page intentionally left blank

Your Free Gift

In order to thank you for your purchase, we're offering a free gift exclusively for the readers of **Dialectical Behavior Therapy: Master Your Emotional Intelligence with DBT, Control Borderline Personality Disorder and Overcome Depression, Anger and Panic Attacks**
By downloading this brochure you will understand better some concepts on this topic.
Click here to access your free gift

Did you know that can download the audiobook version of this book for free?
Click here for Audible US
Click here for Audible UK
Click here for Audible FR
Click here for Audible DE

Don't forget to leave a review on this book. It is simple! Just click on this LINK and you will be directed to the right page. It's very important for me. I'll appreciate if you do.

Introduction

A focal argument among acknowledgment and change lies at the core of DBT. In creating DBT, Linehan first endeavored to apply behavioral hypothesis and change techniques to customers giving BPD and self-destructive conduct. She encountered a few troubles in these beginning times of treatment improvement. Customers' were much of the time non-shared in-meeting, didn't rehearse concurred schoolwork assignments, and frequently didn't return for resulting treatment meetings by any stretch of the imagination.

Linehan estimated that these 'therapy-meddling practices' emerged because the customers saw a solid spotlight on evolving feelings, considerations, and practices as negating. To be sure, as customers frequently accept, they are unequipped for change, the entire idea of treatment dependent on change is essentially negating. Considering these worries, she looked for a way of thinking/hypothetical approach that unequivocally accentuated acknowledgment. Zen standards and practice support the acknowledgment based parts of DBT. To house these two differentiating approaches, Linehan utilizes a powerful way of thinking. The accompanying areas of this paper examine these three establishments of the treatment in more detail.

Pushing for Change: Dialectical Behavioral Theory and Problem-Solving

DBT, like 'first wave' cognitive-behavioral medications, accentuates behavioral hypothesis, as opposed to cognitive hypothesis basic to second wave medicines, for example, Cognitive Therapy for sorrow (Hayes, Follette, and Linehan, 2004). Like 'first wave' therapies, 'third wave' therapies, of which DBT was maybe one of the principals, take an extreme behaviorist point of view to the psychological marvel. Consequently, any reaction of a life form, for example, thinking, emoting, detecting, establishes conduct. The accentuation on the behavioral hypothesis in DBT impacts the treatment approach to the conclusion and case conceptualization.

Predictable with an extreme behaviorist position, DBT sees the indicative criteria of BPD as basic portrayals of the plain and undercover practices of the customer and, critically, that when these practices stop, the conclusion stops to exist. Surely, to an extreme behaviorist:
'A Skill or character is the best-case scenario, a collection of conduct granted by a sorted out arrangement of possibilities' (Skinner, 1974, Chapter 4).

This approach stands out from other hypothetical models of character and character issues that consider the indicative criteria as manifestations of a fundamental 'marginal character' association. A behavioral approach to finding gives an increasingly cheerful point of view to customers. In pre-treatment, DBT specialists depict the behavioral comprehension of the determination, recognize behavioral focuses for treatment, and portray and exhibit how DBT conveys behavioral change. Sketching out that changing both their unmistakable and secret practices evacuates the analysis and situates, customers, towards recuperation.
DBT stresses traditional and operant molding in the event of conceptualization. DBT advisors direct behavioral

examinations to grasp both the traditionally molded connections in the chain of occasions, paving the way to tricky conduct and the useful (operant) outcomes of the conduct. For instance, a customer with a background marked by youth sexual maltreatment every now and again experienced increments in blame and self-destructive ideation while getting ready for bed. Investigation of the increments in ideation uncovered a traditionally molded relationship between hitting the sack and musings of suicide. The customer took in this relationship in adolescence, as the culprit would reveal to her, she had the right beyond the oppressive scenes, which happened in her bed, for which she encountered extreme blame. In the present, after the increments in self-destructive ideation, the customer would look for self-hurt opportunities.

As she looked, she got help from blame as she presently accepted that she was doing 'the correct thing.' Significant alleviation from both blame and self-destructive ideation happened when the customer followed up on her self-hurt or self-destructive inclinations. The customer and advisor recognized that the alleviation from the negative impact and the self-destructive contemplations adversely fortified self-destructive and self-hurting activities, while the conviction that the customer was presently making the wisest decision, 'what was correct', emphatically strengthened these equivalent practices. Behavioral investigations empower customers and specialists to comprehend what triggers and keep up hazardous practices, and in this manner, they structure the initial phase in critical thinking, the center arrangement of progressive techniques in DBT.

DBT advisors utilize the conceptualization obtained from behavioral investigations to create exhaustive arrangement examinations. DBT utilizes standard cognitive-behavioral critical thinking strategies, yet with some novel turns (Linehan, 1993a; Swales and Heard, in press), to diminish tricky practices and increase the securing, fortifying, and speculation of new progressively skillful practices. The advisor, helps the customer to gain new practices yet additionally investigates and comprehends persuasive components that meddle with the usage of new practices. In creating arrangement investigations, DBT specialists utilize four arrangements of progress systems from the cognitive-behavioral ordinance: aptitudes preparing, introduction, possibility of the board, and cognitive alteration.

During the procedure of rehashed behavioral and arrangement investigations, DBT specialists figure out which of these four methods will convey the most extreme advantage to the customer in halting hazardous practices and forming new progressively utilitarian practices. On the off chance that the specialist recognizes that the customer has an aptitude deficiency, for instance, if the customer doesn't have the foggiest idea of how to be properly self-assured, at that point, the advisor will show the pertinent customer abilities. On the off chance that the customer possesses the applicable abilities yet outlandish feelings or useless comprehensions repress the customer from utilizing them, the specialist will utilize presentation and cognitive adjustment separately to enhance the trouble. For instance, a customer may have statement aptitudes; but not use them since they experience overpowering nervousness or believe 'I'm an awful individual for requesting what I need.'

The specialist right now shows uneasiness as the executive's strategies, the cognitive rebuilding of the judgment that requesting what you need is 'terrible' joined with a presentation to making proper solicitations. In the event that the skillful conduct is excessively low in the reaction order, at that point, the specialist will utilize the possibility of the board methods. For instance, the customer may realize how to request what they need; however, both past and current environments rebuff such demands. Right now, the specialist energizes and fortifies the customer requesting what they need, enables the customer to discover environments that fortify solicitations for help, and mentors the customer in how to oversee environments that rebuff demands for help.

DBT specialists create, assess, and execute far-reaching arrangement examinations utilizing the full scope of methodology to hazardous reactions in the behavioral investigation. For instance, in the circumstance of the customer encountering expanded self-destructive ideation on preparing for bed as portrayed before, the specialist utilized a few methodologies. To diminish the traditionally molded increments in self-destructive ideation and blame its happenings on getting ready for bed, the specialist-led imaginal introduction to the sleep time grouping in the meeting. For the customer to encounter a non-fortified introduction during this intercession, the specialist originally practiced with the customer a portion of the care abilities that the customer had learned in aptitudes gathering. An increasingly definite investigation of the sleep time routine uncovered that the customer would, in general, review past upsetting occasions to foresee an expansion in self-destructive intuition as she prepared for bed. The advisor urged the customer to stay aware of the present minute by depicting, in detail, her present activities in getting ready for bed and to just note nosy contemplations about the past or stress musings over the future, in the event that they happened, before pulling together on the present.

During the introduction, the advisor stayed alarmed to the customer getting oblivious and trained her on pulling together on the present. Following presentation during meetings, the customer is working on staying increasingly careful at home while planning for bed. At the point when the customer started to utilize these new abilities at home at night, she called her advisor for extra instructing in the utilization of the aptitudes in vivo (see the segment on Treatment Structure). To address the practical outcomes of the conduct, specialist and customer focused on arrangements both to diminish blame and self-destructive ideation and to expand a feeling of 'making the wisest decision.' The cognitive rebuilding of contemplations of self - fault for the maltreatment demonstrated was viable in lessening blame. To diminish the self-destructive ideation and to build her feeling of 'making the best choice,' the customer helped herself to remember the negative outcomes to herself and her group of self-hurt and inspected her DBT abilities manual to know the skill to use during the present emergency. As she rehearsed the picked abilities, she rehashed to herself, 'Presently, I truly am giving a valiant effort for my family and me'.

Adjusting the behavioral spotlight on change, DBT unequivocally underlines acknowledgment. Linehan drew on her insight into Zen standards to illuminate the utilization regarding acknowledgment in the treatment. Zen standards perceive the flawlessness of every minute, as every minute is brought about by all that went before it, and when proved unable, thus, be in any case or more immaculate than it is (Aitken, 1982; Swales and Heard, in press). Acknowledgment with regards to Zen infers an affirmation of what is instead of endorsement or understanding. The act of approval inside DBT draws on both this feeling of acknowledgment and the acknowledgment of the flawlessness of every minute. The customer is impeccable as the individual seems to be, so is the specialist, just like the connection between them – for how the customer, the advisor, and the relationship could be something besides they are given every one that has happened preceding this minute. Zen standards likewise advise two huge perspectives regarding Zen practice inside DBT: care and radical acknowledgment. Every one of these parts of the treatment will be viewed further.

In approving the customer, the DBT specialist tries to discover reality, knowledge, and precision in the customer's reactions and to feature these. Customers with a BPD analysis have long accounts, and regularly current real factors, of nullification where everyone around them has depicted their convictions, feelings, inward encounters, and practices as improper. Therefore, customers may encounter disarray about which parts of their reactions are substantial and authentic in any one setting. Focusing on which parts of the customers' practices, feelings, and musings bode well empowers them to start to acknowledge their reactions and at last themselves.

Approval assists customers with enduring the outrageous trouble of progress. Swann's Self-verification Theory

(Swann, Stein-Serussi, and Giesler, 1992) underpins Linehan's initial conceptualization of approval (Linehan, 1993a). Swann theorizes that excitement results when people get input conflicting with their self-development. For certain people, irregularities between input got, and self-builds may prompt incredibly significant levels of excitement. Within sight of elevated levels of excitement, the customer strives to recover enthusiastic control, bringing about less coordinated effort, and turns out to be less ready to learn, for example, change. The remedial test for customers with a marginal determination in therapy with a solid spotlight on change is that the therapy negates their confidence in their inadequacy to change. In this manner, at whatever point the advisor endeavors to assist the customer with changing the customer's excitement expansion, their ability to learn diminishes and non-joint effort increases. Given this test, the specialist must titrate pushing for change with the approval of both the trouble of progress and the justifiable doubt in the chance of progress.

Since the production of the treatment manual, Linehan's conceptualization of approval has grown fundamentally (Line han, 1997). In her amended definition, she depicts six degrees of verbal approval and presents the idea of utilitarian approval. Verbal approval includes basically saying to the customer that their reactions bode well somehow or another. The initial four degrees of verbal approval (unprejudiced tuning in and watching, precise reflection, articulating unverbalized musings and feelings, and approval regarding past learning or natural brokenness) are basic in numerous psychotherapeutic models. The two more elevated levels of verbal (approval regarding present setting and radical validity) in spite of the fact that they are not really novel to DBT, is profoundly normal for the treatment. For instance, customers every now and again report that self - harm conduct diminishes nervousness, abstract strain, or other negative feelings states. Right now, injury is legitimate if the customer will likely decrease nervousness. So a DBT advisor confronted with a customer who has cut herself may state, 'It sounds good to me that you cut yourself. This is the main way you know to diminish your tension, and a great many people in a comparable circumstance would need to get their uneasiness down' (current setting approval).

The DBT specialist would likewise push for change. For instance, the specialist may state, 'We have to chip away at different ways for you to get your tension down; however, as the cutting has genuine negative ramifications for you.' Right now specialist negates the invalid parts of the conduct. For instance, with a customer whose objectives are to improve her relationship with her life partner and to prepare as a medical caretaker, proceeding to hurt herself is an invalid conduct corresponding to these objectives.

Radical validity portrays a method for reacting to the customer as the advisor would react to any other individual in their life; for example, the specialist doesn't regard the customer as delicate. For instance, a customer came back to the therapy room and apologized hesitantly for raging out of the meeting and promising not to return. The specialist stated, 'You're correct, that was not your sparkling moment.' The customer looked noticeably diminished at the reaction as she realized that her conduct of raging out was an issue, and the advisor's reaction affirmed her own reaction to her conduct.

The customer, at that point, gave an increasingly revolting expression of remorse, to which the specialist reacted with further approval, 'I'm happy you returned to take a shot at it as I probably am aware change is hard for you.' This model shows the contrast between approval and offering positive remarks about the customer. Approval requires the advisor to check or sanction the exactness of the customer's self-discernment, conduct, or experience in any event when these are negative. Such reactions (paying little mind to valence) may not be simple for the customer to hear; however, they increase the customer's ability to acknowledge and get herself and furthermore, can expand trust in the advisor.

Mastering Emotional Regulation
One of the most noticeably terrible issues we get involved with, inside social settings, is the wonder of feeling controlled by others. At the point when we feel crazy, with the suggestion that others are more in control than we are, emotional regulation turns into a significant test.

Our greatest mix-up in these circumstances is that we expand ideas of control past what we can control. We dive into everyday issues that are no worry to us. It is no big surprise we battle since we have set ourselves in shaky emotional positions.

Mastering emotional regulation is basically about teaching ourselves to concentrate on what we can do and to quit concentrating on what we can't do.
"WHAT CAN I DO?"
At the point when we feel emotionally astray, or essentially somewhat discontent with a specific result, we can pose

inquiries like this, which focus us. For a minute, we might be baffled with the appropriate response. We will most likely be unable to do much with the exception of tolerating the circumstance, all things considered.

In any case, there is an incongruity inside the control here. It's not what the other individual or individuals did or does that matters. However, it's what we did and what we can do that has the greatest effect with respect to the regulation of our feelings.

We assume a lot of liability for others' feelings and insufficient duty regarding our own feelings. No big surprise, we are mistaken for respect to our effect. To stress individual control, we have to focus back again upon what we can impact.

At whatever point we acknowledge the limits inside our very own control, emotional regulation turns out to be rarely simpler. It likewise encourages us to see parts of their brokenness and our completeness.

THEIR BROKENNESS AND OUR WHOLENESS

We repeatedly observe it in a different way; completeness and advantage and our brokenness and absence of limit.

Seeing someone else's unsteadiness is a gift to them, as our elegance excuses them for their frailties of character. Like us, they are a long way from great. They may get things done to disturb us, yet they have less control over their cooperation with us than they might want. What's more, there's nothing amiss with seeing ourselves with a limit with regards to completeness.

By the chance that we are broken, and we are, we additionally have parts of completeness that should be praised.

The limits of our own control are security for us in regard to our emotional regulation. The more we ask, "What would I be able to do?" the more we fortify what we can really impact. Our feelings become less like an exciting ride and increasingly like a consistent drive through the open country.

Dialectical Conduct Measures (DBT) therapy is a type of psychotherapy - or discussion of measures - that uses a cognitive-behavioral approach. DBT emphasizes the psychosocial elements of treatment. The theory behind the strategy is that some people tend to react to certain emotional situations in a more intense and outward manner, particularly as they are situated in romantic, domestic, and PAL relationships. DBT theory suggests that in such situations some people's stimulation levels may increase more rapidly than the normal person, elicit a higher level of emotional arousal, and can take a significant amount of time to return to baseline stimulation levels. Those who are identified on this occasion with a visit to Borderline Personality, swing heavily in their emotions, see the world in black and white and constantly jump from crisis to crisis. Because some humans accept such reactions - all of their own family and childhoods are invalid, they insist - they have no way of coping with these sudden, excessive emotions. DBT is an approach for instructing capabilities that will help in this task.

Components of DBT

Support-oriented: It helps a person find their strengths and builds on them so that the character feels more about him/herself and their lives.

Cognitive-based: DBT helps bring out ideas, and beliefs that make survival difficult: "I need to be the best at everything." "If I get angry, I'm a terrible person" and helps people research different ways. Questioning what would make life more difficult: "I don't need to be the best for humans to take care of me", "Everyone gets angry, it's an everyday feeling."

Collaborative: It requires regular attention to the relationship between employers and employees. In DBT, humans are motivated to overcome problems in their relationship with their physician and therapist. DBT asks people to practice abilities such as whole homework assignments, new ways of interacting with others, and making themselves happy when upset. These skills, an integral stage of DBT, are taught in weekly lectures, reviewed in weekly homework groups, and referenced in almost every group. The personal therapist helps the person learn, practice and understand DBT skills.

Typically, dialectical conduct therapy (DBT) can be seen as having two fundamental components:

1. Individual weekly psychotherapy classes that emphasize problem-solving behavior for the past week's troubles that arose in a person's life. Self-injuring and suicidal behaviors take first priority, observed in a manner of behavior that may intrude on the remedy process. Work can be done to improve the quality of life problems and life in generics. Individual classes in DBT additionally focus on coping with and dealing with post-traumatic stress responses.

2. Weekly Crew Measure Session, usually 2 1/2 hours a session through an educated DBT therapist. In these weekly crew measure sessions, people examine proficiency from one of four different modules: interpersonal effectiveness, crisis tolerance/reality acceptance skills, emotion regulation, and mindfulness competencies taught.

Four modules of dialectical behavior therapy

1. Manmanabhav

Abilities: The essential stages of all abilities taught in the crew are core Mindfulness skills.

Observe, describe and participate, which are the main "skills". They answer the question, "What do I do to practice core mindfulness skills?"

There are non-judgmental, one-minded, and effectively "how" abilities and answers the question, "How do I practice core mindfulness skills?"

2. Interpersonal efficacy

Interpersonal feedback patterns — how you interact with people and in your personal relationships — that are taught in training DBT abilities, share some assertiveness and similarities with those taught in interpersonal problem-solving classes. These skills incorporate positive techniques that ask how to learn, to inevitably face interpersonal conflict.

People with borderline personality illness often have accurate interpersonal skills. They travel, however, in the application of these skills in specific contexts - particularly in emotionally susceptible or risky situations. When someone is facing a difficult situation with another person, they may be able to describe spectacular behavioral scenes, but when examining their personal situation, it is possible to generate or exclude the same set of behaviors. As they may be unable to carry at all.

This module focuses on situations where the goal is to change something (e.g., to ask someone to do something) or to try to make someone else cope with the adjustment (e.g., not announcing what to do). The skill taught is to maximize the chances that an individual's goals are met in an exact situation, while not damaging the relationship or the person's self-esteem at the same time.

3. Tolerance Most methods of mental health therapy focus on risky activities and changing circumstances. They have given little interest to accept, search for meaning, and endure the crisis. This mission has usually been dealt with through non-secular and religious communities and leaders. Dialectical conduct therapy emphasizes knowing how to efficiently tolerate pain.

Ability to bear the crisis, skillfully represents an herbal improvement. They have to do with the ability to accept each individual and contemporary position, in a non-evaluative and non-discretionary fashion. Although the stance here is a non-controversial one, it does not mean that it is one of approval: accepting reality is not an acceptance of reality.

Crisis tolerance is related to behavioral tolerance and living crises and accepting existence as it is in the moment. Four units of crisis avoidance techniques are taught: distracting, self-soothing, improving the moment, and thinking of pros and cons. Acceptance skills include radical acceptance, turning the idea toward acceptance, and will versus desire.

4. Emotion Regulation

People with borderline personality disease or who may be additionally suicidal are usually emotionally intense and inflammatory - often angry, acutely depressed, depressed, and anxious. This suggests that humans struggling with these concerns will likely benefit from helping them acquire knowledge to adjust their emotions.

Dialectical conduct therapy capabilities for emotion legislation include:

Learning how to get out and label emotions

Identifying boundaries for changing emotions

Reducing vulnerability to "spirit mind"

Increase in good emotional events

Chapter 1.

The Theory and Research Behind DBT.

Dialectical Behavior Therapy (DBT) 1 is advanced by Marsha Linehan's efforts to create a treatment for multiparous, suicidal women. Linehan combed through the literature on effective psychosocial treatments for various disorders, such as anxiety disorders, depression, and other emotion-related difficulties, and immediately assembled a bundle of evidence-based, cognitive-behavioral interventions involving suicidal behavior. Initially, these interventions were so focused on changes in cognition and behavior that many victims were criticized, misunderstood, and held invalid, and consequently dropped out of treatment altogether.

Through the interaction of science and behavior, multidisciplinary, medical experiences with suicide victims promoted research and medical development. Most notably, Linehan weaved into medical interventions designed to bring patient acceptance and help the affected person to give himself, his feelings, and thoughts, to the world and others. As such, DBT got here to rest on the basis of dialectical philosophy, whereby practitioners continually try to synthesize stability and acceptance and change-oriented strategies.

Ultimately, the work culminated in a comprehensive, evidence-based, cognitive-behavioral treatment for borderline personality disorder (BPO). The fashionable

DBT remedy package includes weekly in-person medical classes (approximately 1 hour), a weekly crew efficiency education session (approximately 1.5–2.5 hours), and a physician counseling team assembly (approximately 1–2 hours). Currently, eight published, well-controlled, randomized, scientific trials (RCTs) have confirmed that DPT is an effective and specific treatment for BPD and related problems.

This article sheds light on several key elements of DBT and answers central questions that can also help clinicians to find out whether to implement treatment. In doing so, this article specifically sheds light on aspects of the theory and practice of DBT, distinguishing this treatment from different perspectives as appropriately affected person populations, and integral and unique elements of DBT that must be in the area for any patient.

In order to know whether to use DBT or different preventions for a patient, there is an important piece of research about the treatment of the victim, which is similar to, or diagnoses, the affected person's case. It tells the person about the characteristics. Researchers and remedy builders have used DBT to the extent of the affected person's population, although the prevalence of RCT has targeted people (mainly women) with BPD.

The following section contains a brief overview of well-controlled RCTs that have evaluated DBT.

Parasuicide patients with BPD: For parasuicidal BPD patients, the most stable finding is that contrast with control conditions on DBT cuts the best rate of parasuicidal behavior. The first RCT of DBT (n = 44 parasiticidal females with BPD) determined that DBT improved treatment management status as it typically occurs in community (TAO, or treatment-as-usual) frequency. And occurs in reducing clinical severity, parasuicide, instant hospitalization days, trace anger, and social functioning. Through the first six months of the 12-month follow-up period, patients with DBT showed less parasuicidal conduct and anger and better social adjustment. Findings related to higher social adjustment during the last six months of the follow-up period, revealed that DBT patients additionally had fewer inconsistent mental days during this period.

The most recent and largest RCT of DBT (n = 101) studies with more rigorous management status consisting of treatment through neighborhood physicians as experts in the treatment of BPD (treatment-by-community specialists, or TBCE) repeated this study and determined that suicide attempts in DBT patients, psychiatric hospitalization, the scientific danger of parasitic behavior, angry behavior, and emergency room visits, in contrast to TBCE patient had 5 during 12 months of therapy and 12 months Follow-up behavior at a large discount rate. Period.

A couple of researches have investigated DBT for girls with BPD in neighborhood settings, such as the Community Intellectual Health Corps and the VA Hospital. In a neighborhood mental health setting, Turner 6 compared a modified model of DBT to a fully individualized measure of client-centered medical control status. Patients with DBT conditions had major scars in suicide attempts, intentional harm to themselves, uncomfortable days, suicidal tendencies, impulsivity, anger, and global intellectual health problems. In addition, finding out about female veterans with BPD suggests that DBT victims had larger symptoms of suicidal tendencies, hopelessness, depression, and anger than did Tau patients for these two types of research. Follow-up data for both these researches are no longer available.

Women with BPD and substance use disorders.
The second patient group for which DBT has tested promising records is women with BPD and disease-using substances (SUDs). The first study at this location compared DBT to Tau for a woman meeting the criteria of BPD and SUD8 and found that DBT victims underwent 12 months of therapy and adherence to a four-month period of drug use. They showed an increase in discount rates and a reduction in drop out fees during treatment. Regarding a discovery made through Linehan's group for 2D, opium-dependent women with BPD were randomly assigned to two conditions: DBT or a rigid management position, called the 12-step (CVT-12S) which is known as a comprehensive validity treatment. In both conditions, participants additionally acquired LAAM acetate hydrochloride, an opium alternative drug. The CVT-12S includes a stripped-down version of the DBT that is the only related acceptance-oriented intervention designed to manipulate access time to treatment, tutorial therapy settings, and therapist experience and commitment. Participants in both DBT and CVT-12S confirmed a substantial remission in opium use during the 12 months of treatment, but DBT patients had increased continued abstinence from opium use at 16 months of follow-up.

A couple of RCTs conducted outside the US have also investigated DBT for drug abuse with BPD. Conducted in the Center for Addiction and Mental Health (CAMH) in

Canada in contrast to the popular DBT for Confirmation of Common (TAU) and One Substance Use Disease (N = 27) .10 DBT patients as a treatment for women with BPD. There is an increase in suicidal and parasuicidal behavior and alcohol use, but not in the use of separate medication.

A search conducted in the Netherlands 11,12 included BPD patients, 53 percent of whom met the criteria for substance use (SUD). The findings indicated that DABT sufferers showed greater declines in parasuicide behavior and impulse-control problem behaviors (including binge, gambling, and reckless driving, but not substance abuse) than TAU patients. DBT patients were forbidden to exhibit low parasuicidal behavior, impulsive behavior, and alcohol use at some point in the six-month follow-up period.

Other scientific populations and problems
Additionally, some research has investigated DBT-oriented therapies for other clinical problems, as well as increased intake of disorders and sadness in older patients. Telch and colleagues compared the 20-week DBT-based abilities coaching group with the waitlist control condition for women with the binge-eating disease and found that in a binge, body image, conflict concerns, and anger in DBT patients, there was a big upgrade. Although 86 percent of DBT members had stopped circling during treatment, the number had dropped to 56 percent during the six-month follow-up period. A 2D study in contrast to a modified model of individual DBT that blankets skills training for a waiting list condition showed that DBT patients had higher remission rates in binge and purification. No follow-up facts are at hand for this latter study.

To find out about depressive elderly victims who met the criteria for a personality disorder, there was only 15 investigators for medications, as opposed to a customized version of DBT plus antidepressant medications. The findings indicated that a large proportion of DBT sufferers had relieved depression after treatment and over a six-month follow-up period.

Summary

In summary, sufferers for whom DBT has the strongest and most routine empirical assistance include pericyclic lids with BPD. There are also some promising records on DBT for women with BPD who fight substance use problems. Preliminary information proposes that DBT may also promise to reduce binge-eating and other eating disorders. On the one hand, the most conservative clinical option would be to restrict DBT to a woman with BPD. On the other hand, DBT is a comprehensive therapy that includes various evidence-based, cognitive-behavioral intervention factors for other clinical problems. As such, DBT is often applied in general settings to polyandry victims in general, victims who have comorbid axis I and II disorders, and/or who are suicidal or self-injured; However, caution is necessary for using a therapy for patients with whom it has been evaluated in research.

Go for:
Important and unique elements of DBT
The following area contains a dialogue of some important and unique factors of DBT. DBT is a comprehensive treatment that incorporates several elements of various cognitive-behavioral approaches, such as to conduct therapy (i.e., exposure, contingency management, problem solving and stimulus control), cognitive restructuring, and

other such changes. Many of these interventions are much greater than those found in other treatments, the emphasis here being on the indispensable components of treatment that are special and exclusive to DBT, including (a) the five functions of treatment, (b) Focus on emotions in quadratic theory and treatment, (c) dialectical philosophy, and (d) acceptance and mindfulness.

Five features of treatment
DBT is a complete application of therapy consisting of individual therapy, group therapy, and a therapist session team. In this way, DBT is a software of treatment, rather than a single treatment approach performed using a physician in isolation. Often, doctors are interested in using DBT but explore the possibility of implementing such a complete measure to be challenging. In this case, it is important to know whether the most essential element of any DBT program is whether it addresses the five key features of treatment. Although DBT's fashionable bundle has the most empirical support, different settings and opportunities may require modern and creative applications of DBT. However, in all cases, it is necessary that any optimization of DBT performs the following five functions:

Function # 1: Enhancing capabilities.
Within DBT, the notion that people with BPD both significantly decrease or improve essential lifestyle skills includes (a) controlling emotions (emotion formation skills), (b) current times, to pay attention to the journey and regulating interest (skills), (c) efficiently navigating interpersonal effects, and (d) enduring and avoiding crises (crisis tolerance skills) apart from enduring and avoiding crises. Doing is the key to increased capabilities. The

ceremony is typically performed through a weekly abilities group session, involving approximately 4 to 10 people, as well as homework assignments to assist patients' practice skills between sessions.

Function # 2: Generalizing capabilities.
If the abilities felt in the treatment sessions no longer switch on patients' lives every day, then it would be difficult to say that this remedy was once successful. Consequently, a second unavoidable feature of DBT involves normalizing healing beneficial properties in the patient's natural environment. This feature is executed in skills training using homework assignments for practice and problem-solving on how to improve the practice of abilities. In male or female therapy sessions, therapists help victims to follow new competencies in their everyday life and often patients have a habit of behaving efficiently in sessions. In addition, therapists are accessible with the help of cell phones between classes to observe the abilities of affected people when they most desire (e.g., in a crisis).

Function # 3: Improve motivation and reduce dysfunctional behavior.
A 0.33 feature of DBT emphasizes improving patients' motivation to change and reduce behaviors inconsistent with lifestyles for patients to live. This ceremony is mainly achieved in character therapy. Each week, the therapist completes a self-monitoring shape (called a "diary card") on which he or she tracks more than some treatment ambitions (e.g., self-harm, suicide attempts, emotional grief). The therapist uses this diary card to prioritize the time of the session, giving the behavior that most prioritizes the patient's survival (e.g., suicidal or self-

injuring behavior), the behaviors that follow through (e.g., absenteeism, latency, non-behavioral behaviors), and behaviors that result in the patient's best life (e.g. extreme problems in living, unemployed) interference with severe problems associated with the disease.

After prioritizing behavioral ambitions for a given session, the therapist helps the affected person find out what was done for the behavior(s) in the query and reinforce or contain the behavior(s). Penalties can be applied. The therapist additionally helps the patient to find ways to be efficient, behave well, clarify issues in life, or change feelings. In motivation-enhancing phrases, the therapist actively works to get the person committed to change, using a series of "commitment" strategies.

Function # 4: Enhancing and preserving physician abilities and motivation.
Another essential function of DBT is to maintain the motivation and abilities of physicians dealing with patients with BPD. While assisting BPD patients can be stimulating and rewarding, they also engage in a powerful combination of victim behaviors that can tax coping resources, competencies, and their remedies to proffer solutions (i.e., suicide efforts, repeated suicidal distress, interfere with behavior therapy). Consequently, an essential component of excellent treatment for BPD sufferers is a system of providing assistance, validation, continuous training and skill-building, feedback, and encouragement to the physician.

To deal with this task, the comprehensive DBT consists of a meeting of a physician consultation team, for which DBT practitioners meet for about 1 to 2 hours per week. The team proffers ways to help take beneficial measures, especially in the face of medical challenges (e.g., a suicidal patient, a patient who misses sessions). In addition, the crew encourages physicians to maintain a compassionate, non-controversial orientation towards their patients.

Function # 5: Structuring the Environment.
A fifth important feature of DBT involves structuring the environment in a way that reinforces fine behavior/progress and no longer behaves malicious or easy. Often, it structures treatment in a way that promotes progress correctly. Typically, in DBT, the character physician is the primary physician and "in charge" of the medical team. He or she makes sure that all the factors of high-quality treatment are in one place, and all these functions are met.

Structuring the environment may additionally involve supporting victims in finding ways to regulate their environments. For example, patients using the drug may additionally need to know how to adjust or keep away from social circles that promote drug use; Sufferers who are once suicidal want to learn how to ensure that their partner or others do not commit self-harm (i.e., with the help of being highly soothing, warm, or supportive). In DBT, the physician typically has the patient change his or her environment, but sometimes, it may take an energetic state to change the patient's environment (for example, if the environment is heavy or if the patient is very powerful).

Emphasizing ideas in biosocial thought and treatment

In addition to serving the five characteristics described earlier, DBT is anchored in a theory of BPD that motivates clinicians to a center point on emotion and emotion regulation in treatment. According to BPD's Biosocial Theory, people with BPD are born with a biologically difficult-tempered temperament or temperament for emotionally difficult. The emotional vulnerability has a very low threshold for emotional arousal, severe emotional reactions, and withdrawal for difficulty. Baseline stages of emotional arousal. Without very efficient and superb parenting or child upbringing, the child has difficulty mastering or coping with such extreme emotional reactions.

The central environmental factor consists of a posterior environment that invalidates the child's emotional reactions by ignoring, dismissing or punishing them, or simply using a method of mimicking /problem-solving. Invalid environment behaves sensitively to the child's emotions, thus increasing the risk of developing BPD. As a result, the infant goes beyond the intending abilities to modify emotion, often fearing its own feelings (i.e., "emotion phobic"), and a soon-to-be-executable, self-destructive approach to coping with it can take the support of feelings (for example, self-harm).

DBT is an emotion-focused treatment based on the sensitization of BPD as an emotional disorder. One of the essential goals of DBT is to help make patients' survival enjoyable, with a reduction in "pleasant to exist" ... ineffective action disposals associated with passive feelings. Thus, DBT includes a number of behavioral skills that are primarily intended to instruct patients. Identify, understand, label, and adjust their thoughts (i.e., emotion regulation skills). In the DBT session, the therapist participates in the patient's emotional reactions, especially when they interfere with progress, and many of the most commonly used manuals in DBT helps patients regulate their emotions.

Along these lines, in using DBT to sufferers with BPD, practitioners needed competencies and wanted to work with ideas in knowledge treatment. In particular, the therapist should be knowledgeable about looking at emotions and emotion regulation. Furthermore, various imperative skills for physicians include (a) their role in thoughts and complex behaviors, (b) taking into account the patient's emotional reactions, facial expressions, physiologic language, voice tone, and other such symptoms of emotional states support (C) patients to accurately label emotional states, (D) validate emotional responses or conditions that are statistically healthy, (d) discrimination when unique ability is likely to be beneficial in helping patients adjust (or accept) their feelings, and (e) apply sense rules techniques to educate when emotionally overwhelmed.

Dialectical philosophy in DBT

Dialectical philosophy is gasoline which is very unique about DBT in its evaluation of other cognitive-behavioral therapies. Dialectic philosophy is most commonly associated with the thinking of Marx or Hegel but has existed in one form or another for thousands of years. Within a dialectical framework, opposites, in reality, are polar forces that are under stress. For example, the pressure to follow change-oriented treatment strategies changes the patient's willingness to stress rather than broad-spread through increased pressure. The dialectical philosophy additionally states that every opposing pressure is incomplete in itself and that these forces are constantly balanced and synthesized.

This is additionally the case in DBT. On the one hand, focusing solely on change-oriented efforts was an incomplete strategy, as it lacked a critical component of acceptance. On the other hand, focusing solely on patient acceptance can also be incomplete and ineffective, because as a multicomponent, patients with suicide require considerable modifications to create the lives that are worth living.

Dialectical thinking influences many components of the therapist's strategy and style. For example, the therapist consistently seeks to balance and synthesize acceptance and change-oriented strategies in the most spectacularly feasible way. Within each session, the therapist works to provide consistency of acceptance and validation with alternative strategies of problem solving/behavior. In suggesting solutions or skills, he is often involved in each acceptance-based (e.g., radical acceptance, enduring crisis, being aware of emotional or other experiences of modern times) and change-based (e.g., problem solving) solution. When the therapist and the affected person play the horn on particular issues, dialectical questioning lets the therapist go with a desire to be "right" and pay attention to the ways the person is synthesized with the affected person. Based on the idea that each work is possibly incomplete in itself.

Finally, in DBT, movement, speed, and weight are emphasized within therapy sessions. Therapists use a range of remedy techniques and vary their style and intensity from lively and energetic to slow and orderly, and from interpersonal and legitimate and irrelevant. In addition, practitioners adjust their strategy completely based on what is/is not working at the moment.

Acceptance and thoughtfulness in DBT

In DBT, many interventions and abilities are shown to allow the patient and help the affected person to take delivery of his or herself, others, and the world. One such intervention is mindfulness. In DBT, mindfulness skills help patients to be present. Some of the mindfulness abilities include and are non-incidentally modern to describe the data of the present-day experience or situation and to fully cooperate in the current activity/experience, as well as to participate in one factor at a time, participate in experience, and "one-minded" 16 and focus on effective, efficient behavior. Therapists teach temperament abilities to victims in aptitude training, encourage individual estrangement in therapy, and regularly practice self-contemplation.

As taught in the Crisis Tolerance module of skills training, any other acceptance intervention in DBT is known as radical acceptance, which involves truly accepting the journey of the present moment for what it is, its exchange. To prevent conflict or to deliberately oppose it. Finally, each other acceptance intervention in DBT allows the patient to accept through recognition, including validating or accepting validity or truthfulness in the patient's experience, emotional responses, thoughts, or opinions. A basic skill for physicians in DBT (as mentioned earlier) involves understanding when and how to practice the most amazing acceptance-oriented strategies, given the patient's characteristics and difficulties and the context of the therapy session.

Chapter 2.

How (DBT) Diverges from Traditional Cognitive Behavioral Therapy Approaches.

CBT (cognitive-behavioral therapy) and DBT (dialectical behavior therapy) are two forms of psychotherapy "Talk Therapy." In both, you work with a mental health expert who performs additional studies about the challenges you experience and analyzes to help manipulate abilities and challenges on their own.

Cognitive-behavioral measures or CBT teach you how your thoughts, feelings, and behavior affect each other. For example, if you agree that man is not like you (thought), then you can avoid social situations (behavior) and loneliness (feeling). However, CBT teaches you how to use these relationships to your advantage: a high-quality change in one thing (changing an idea or behavior) can make good modifications to all factors. CBT is an approach that has been validated through research to work towards a number of specific intellectual health problems, such as depression, anxiety disorders, ingestion problems, and substance use problems.

CBT is currently structured, short-term, goal-oriented and targeted. It starts with training around a particular intellectual illness or venture and how the illness or undertaking affects you. Next, you will analyze and practice skills and techniques such as problem-solving or sensible thinking to help bring changes in your thoughts, feelings, and behaviors. You will analyze how you can use your new abilities to deal with problems in the future.

The dialectical behavior measure or DBT is mainly based on CBT, with a larger focus on emotional and social aspects. DBT was once developed to help people cope with extreme or unstable emotions and harmful behaviors. DBT is an evidence-based strategy to help change emotions. It began as a treatment for borderline personality disorder, and contemporary research suggests that it may also help with many different intellectual illnesses or concerns, especially self-harm.

The main variations between CBT and DBT are validation and relationships. DBT teaches you that your experiences are real, and it teaches you how to deliver regardless of challenges or challenging experiences. Relationships in DBT are also very essential - including the relationship between you and your DBT practitioner. You can also have a normal check-in to talk about any successes or problems. Treatment may also include a combination of individual periods and team sessions. In addition to CBT skills, you will use research skills to manage your emotions, build relationships with others, work properly with troubles or distress, acceptance and self-will.

As with many therapies, the benefits of CBT and DBT skills take time and effort. But as soon as people get hold of abilities with the support of their CBT or DBT therapist, they regularly find out that their new abilities and tactics become 2D nature - they are tools that will last a lifetime.

Psychiatry is one of the first-rate medical strategies available for a variety of mental illnesses. One of the most common types of therapy is called cognitive-behavioral therapy (CBT). Also known as talk therapy, CBT focuses on speaking about your troubles to help you understand your thoughts differently. If you feel that terrible thoughts are constantly under management - "I am a failure." I can't do anything right. Nobody will like me if they see who I really am. "- CBT can help you use good judgment and reason to flip the script and be in the manipulation of your thoughts rather than allowing your thoughts to manipulate you.

However, the well-known CBT in treating all intellectual ailments is not spectacular. Another frequent type of therapy is known as dialectical behavior therapy (DBT). DBT is a specialized structure of CBT that focuses on helping humans who have extreme emotional reactions interact with their surroundings in a less emotional, healthier way.

With such comparable names, there are CBTs and DBTs that are distinct from each other, and can the use of one have an advantage over the use of the other? Read on to learn about the difference between CBT and DBT and to find out which one will help you a lot.

CBT is a generic term
CBT is a catch-all phrase for treatment options that share common characteristics. DBT is a type of CBT, along with countless other types. Physicians practicing CBT usually speak of remedies that depend on quite a few guiding characteristics. Those elements include:

Treats emotional response: Primarily based on the assumption that our thoughts have an effect on our emotions, so changing the way we think and react to situations will help us understand better.

Restricted to a specific period of time: Most patients will seek therapy for a period of time and then begin following CBT strategies on their own without the help of a mental health professional. If anger or conduct problems persist, clients may also develop an extraordinary type of therapy to address a particular type of trauma or another issue that causes pain or an obstacle to leading a full life.

Having a good physician-patient relationship: CBT works great when the patient feels they can trust their physician. Due to the non-public nature of medicine, patients need to seek out someone they respect and are satisfied with.

Depends on cognition and justification: CBT encourages patients to observe the rationale and to help direct how they all react to situations as alternatives to the driving thoughts.

The Guide Uses Structure for Treatment: Therapists have a unique purpose for the techniques and techniques they express in each session. They use the client's wishes to parent what CBT standards would be most recommended to them and tailor one according to each.
While DBT is a type of CBT, it is tailor-made to support well-known aches and pains to the human, but still feel safe and "cured" at the moment and impulsive or harmful. It gives the right to make healthy behavior instead of

actions. While some emphasis is given to considering thoughts, patients are taught to search for triggers outside themselves and to fit these triggers with a healthy coping mechanism or response.

CBT vs DBT for treatment of certain diseases

Not all intellectual illnesses respond to treatment in the same way. A medical approach that works well for sadness and anxiety can exacerbate consumption issues or character disorders.

CBT has been shown to be quite amazing when treating depression and has the additional potential to relieve sadness compared to different types of therapy. This approach to therapy has additionally been considered helpful in treating anxiety, as it manipulates victims to recover. CBT has additionally been shown to be helpful in dealing with the obsessive-compulsive disease (OCD), phobias, panic disorder, post-traumatic stress disorder, and sleep problems.

DBT was once created to help people recognized with a borderline character disorder. DBT focuses on supporting people in alternating their conduct patterns, as it is counterproductive to speak or try to speak through the issues they are struggling with. These types of CBTs help those who have developed patterns of extreme emotional reactions and impulsive behaviors that patients describe as extreme feelings of pain and rejection - a feeling of walking through a knife-ridden world. DBT is often the best therapy for those who battle with self-harming behavior, like slicing and frequent suicidal behavior. Sexual trauma survivors additionally respond to DBT techniques.

Darshan CBT vs DBT Used

CBT focuses on logic and justification, as do most of the atonement located in the Stick philosophy and Socratic method. The Socratic method uses compulsory thinking to query the assumptions in place. It works well for those who go through anxiety and depression, as it helps them to see their problems from a more logical factor. For example, customers who fight with feelings of failure and inadequacy are requested to look into the facts. When have they actually succeeded in engaging in a goal? Are there other humans - friends, family or co-workers - who want to testify to the success of the customer in exceptional situations? Who or what do they use as a measuring stick for success? Is this a practical comparison?

DBT relies heavily on mindfulness skills used in Buddhism and Zen practices. DBT teaches sufferers to use precise mindfulness strategies to analyze living with pain in the world, and to achieve how to do things rather than struggle with how to exchange things.

Differences in treatment methods

CBT focuses on how your thoughts, feelings, and demeanor impact every second. While DBT works on these things, the emphasis is towards regulating emotions, being mindful, and acquiring knowledge to accept pain. CBT wants to give patients the ability to understand when their thoughts may turn out to be troublesome, and provide them with techniques to redirect these thoughts. DBT helps victims find ways to self-deliver, is sensibly safe, and helps adjust their carefree adverse or dangerous behaviors to manage their thoughts.

Customers interacting in a DBT measure participate in a coaching period of DBT capabilities typically taught in a team that is put into four modules. Most patients additionally teach DBT therapists or DBT weekly and want to receive DBT phone teaching when they need the most help. Occasionally, as patients are able to use DBT capabilities to modify their emotions, increase temperament, and enhance relationships with others, they are specifically used to address poor thinking patterns or simple harmful behaviors. Additional comprehensive CBTs are able to transition to organizations.

How to tell if CBT or DBT is right for you
The excellent way for patients to decide what type of medical treatment is best for them is to talk with a mental health professional - a therapist, psychiatrist or psychologist. They will think about your symptoms, treatment history, and the desires you want out of therapy, and advocate for high-quality next steps.

What is your diagnosis?
Because each disease responds differently to techniques, you would prefer to go with the technique that has proven to be the most excellent for your diagnosis and treatment of symptoms. If you have not yet received a diagnosis from a psychiatrist or psychologist, consider an appointment for psychiatric disease and psychological testing. This will help you discover great remedy options, as well as strengths and current abilities that you can take advantage of in your healing process.

Depression and anxiety sufferers have had much success with CBT, while humans with borderline personality disorder and chronic thoughts of suicide find DBT more useful. Keep in mind that many humans have more than one diagnosis, and some people use elements from both DBT and CBT to manage their symptoms.

Have you already tried therapy?
Many individuals who have gone to the measure say that they do not want to go back because they feel it was not effective. Because the patient-physician relationship is so important, think about interviewing a few different therapists to see if you can discover a higher match. You may additionally like to consider trying out a specific "taste" of CBT. In addition to DBT, CBT is a complete alphabet soup of various variants of Acceptance and Commitment Therapy (ACT) and Mindfulness-Based Cognitive Therapy (MBCT). If one of the goals of therapy is to improve your relationships with others, then consider couples treatment or family remedy as an alternative to doing it alone.

And provide it a few weeks in advance until it leaves the name. Remember that your doctor is no longer going to work for you. You want to commit to doing the hard work of revising your lifestyle for more healthy thinking, healthy behavior, and healthy living. Recovery does not appear overnight, but by using a medical treatment that matches your symptoms, and searching for the right doctor or psychiatric treatment program, you can gradually create a collection of small modifications that greatly reduce pain and a better life.

The current speed propagation of "new wave" therapies, also loosely known as third-wave cognitive-behavioral therapy, has renewed interest in their comparative performance, scientific validity, and theoretical and methodological integrity. However, critics are additionally expressing concern that these cures are not properly supported and are "getting in advance of the data". This article engages the literature on a small selection of modern medical approaches, namely mode inactivation therapy (MDT), acceptance and commitment therapy (ACT), and contrastive behavior therapy (DBT), and cognitive-behavioral therapy (CBT), more specifically.

On the one hand, it challenges the notion that in terms of MDTs, ACTs, and DBTs in general, an additional 0.33 waves are fine, much less new to provide through the most desirable performance, often a population context. Common techniques like CBT are generally considered to be difficult to treat. In different ways, however, it claims that 1/3 wave treatments are alternatively more part of the CBT family of procedures than different treatments. It draws itself from the concepts, practices, and experiences of related developers that argue for the amazing cost and potential that these 0.33 wave treatment processes provide in phrases of scientifically established remedy outcomes. MDT, for example, is difficult and unintelligible to dismiss, with the most promising outcomes in the treatment of severe multi-problem adolescents, a population that has an exceptionally high human and economic impact in the long run. By a better understanding and differentiation of third-wave therapy, their spread can be accelerated without compromising their manageable extra efficiency and focus.

SO-CEDED THIRD WAVE Therapeutics, Dialectical Behavior Therapy (DBT) and Mode Deactivation Therapy (MDT) along with Acceptance and Commitment Therapy (ACT), are results of Cognitive Behavioral Therapy (CBT). (Classical behavioral therapies are referred to as the first wave and the classical cognitive therapies as the second

wave.) These three types of measures are currently showing an expanded amount of success with adolescent childhood behavior disorders struggling with issues like Post-traumatic stress symptomatology, and other temperament issues (Apps, Dimeo, & Kohlenberg, 2012; Appsche, Bass, & Backlund, 2012; Powers, Wodding, & Mmelkamp, 2009). Dialectical behavior therapy
Using DBT in 1993, Proc. This was done through Marsha Linehan, who aspired to adapt CBT when she identified approach deficiencies with patients with her borderline personality disorder (Bayles, Blossom, & Apsche, 2014).

The essential goal was to accommodate these particular characteristics such as high emotional sensitivity and vulnerability to perceived rejection. DBT uses a variety of CBTs that teach the patient the unique competencies they need to cope with stress and to help regulate their emotions (Apps, 2010). Since then, it has proven useful for treating a large range of issues, including suicidal behavior, substance abuse, spreading disorders, and sadness (Murphy & Siew, 2012; Dimeff & Koerner, 2007).

The basic goal of DBT is to train the character in the skills they will need to cope with stress (Apsche & DiMeo, 2012). This gives them the tools that they will need to transform their modern terrible mimicry mechanisms into brilliant ones that will help them change their views and improve relationships with others around them (Apps, 2010). It will then instruct the man or woman that they need to cope gracefully with stressors, adjust their emotions and help them grow relationships with their parents and various household members. There are four key elements that make DBT successful; (A) Cognitive-behavioral theoretical framework, (B) validation, (C) dialectics, and (D) radical acceptance (Bayles, Blossom, &Apsche, 2014).

DBT uses dialectics which implements the notion that wholeness is made up of opposites and that change occurs when one opposing force is superior to another. It has three basic assumptions, the first being; all matters are interlinked. Second, change is constant and inevitable. Third, the adversary can be constructed to shape an implicit approximation (Apps, 2010; Murphy & Siew; 2012; Upsche, Diemo, & Kohlenberg, 2012). Today DBT is basically used with people who present strong urges to harm themselves and those who have self-destructive behavior (Upsche, Diemo, & Kohlenberg, 2012). This is one of the reasons why it is beneficial in adolescents, especially young men. DBT additionally encourages acceptance and change. Acceptance is unconditional and is emphatically linked to the direct exchange of ideas (Upsche, Diemo, & Kohlenberg, 2012).

DBT has been shown to enhance behavior in three strong areas that disadvantaged adolescents typically seek to improve: (a) a lack of skills that mimics the required behavior, (b) accepting fact as a factor, and (C) to preserve a strong commitment to trade (Arch, Eifert, Davis, Willardaga, Rose, & Kruske, 2012). DBT offers capabilities that aspire to end the productive contributions of external neighborhoods in which people live and helps them get a good outlook on the value and acceptance of their personal struggles, and fine exchange (apps), Sew, and Abilities. (Mateen, 2005).

A downfall to using CBT elements in the DBT methodology is that its nature is in a consistent mission to people's emotions, making it difficult for them to accept their beliefs as genuine and justified. As such, they want to choose that other people do not perceive their emotional pain as real and try to attract attention when, in reality, they actually suffer and experience emotional pain. One great feature of CBT factors in the DBT method is that it provides a way to acknowledge the experience of their actual pain (Apps, Siew, & Materson, 2005). This additionally assures that their modern-day malicious behavior is clearly understandable and shows them that they should trade this way of coping while giving them more high-quality solutions to their problems. Helping discover ways that will be healthier for his wellbeing and thinking set (Bayles, Blossom, & Apsche, 2014).

In addition, DBT uses CBT principles to create a bond between patient and physician so that they can work collectively to get to the bottom of issues and provide new ways for the affected person to accept them getting stressed and alternate in a great way to help them.

Acceptance and Commitment Therapy

With acceptance and commitment therapy (ACT), mode deactivation therapy (MDT), cognitive behavioral therapy (CBT), and dialectical behavior therapy (DBT), all relate to the third wave of cognitive-behavioral therapies (Apps, 2010, Zettle, 2012). The primary purpose of ACT is to help the affected person continue to be aware of their personal memories, thoughts, and feelings so that they need to alternate or avoid them (Apsche, 2010; Ruiz, 2010; Jennings, Apache, Blossom, &Bayles; 2013). Like MDT and DBT, ACT additionally uses mindfulness techniques to regain acceptance, dedication, and positive behavior change. As a result, CBT uses components that focus on cognition, and then allows adolescents to shift the focus of attention by challenging or controlling them as an alternative to accepting bad thoughts or feelings (Jennings, Ash, Blossom, and Bayles 2013). The ACT then isolates negative thoughts and behaviors by allowing teens to exchange brilliant ideas with the use of mindfulness theories.

Mindfulness in the ACT is the building bridge between the collaborative relationship between the affected person and the therapist (Jennings, Apsche, Blossom, & Bayles, 2013). Mindfulness in the ACT is used to defuse language, raise awareness, and understand oneself as an observer of thoughts, feelings, and experiences (Romanoff, 2012). Committed to taking action in return, this therapy additionally helps adolescents. The ACT has used diffusion to separate emotions from words (Ruiz, 2010). In the ACT, cognitive illusion strategies are used as an alternative to denying their existence in an attempt to change the undesirable features of ideas and other nonpublic events.

Whereby unpleasant thoughts and feelings are deployed in their appropriate context where their influence is minimized. Patients are encouraged to faithfully and impartially observe their thoughts and opinions, except to engage them as their own integrated components. In short, ACT helps patients "connect with a self-permeating, compassionate feeling that is superior to our constructed identity or self-created concepts (Romanoff, 2012, p. 134).

The use of the philosophy of functional contextualism in ACT helps it to analyze specific behaviors that rude behavior in phrases of its particular feature in a very discrete context (Arch, Eiffert, Davis, Willardaga, Rose, &

Craske, 2012) are causing. In purposeful contextualism, psychological events — thoughts, feelings, and behaviors — are used over time by focusing on manipulative variables in their particular context. Thus, pragmatic contextualism has been developed to aid in elucidating psychological problems that remain consistently bleak or ambiguous, although at the core of the individual's complex conduct (Romanoff, 2012). It is used through the ability to predict and influence. In the ACT paradigm, the patient's thoughts and feelings are no longer seen as being right or wrong, but rather in an appreciation and changing existence beneficial for the journey of greater duty and happiness (Ruiz, 2010).

Mode passive therapy
Mode Deactivation Therapy (MDT) is one of the most current types of therapy in treating aggression and habitual illness among adolescent girls. The measure was developed in the late 1990s with the help of Dr. Jack Ackes to overcome the barriers to CBT with grizzled multi-problem youth who have habitual disorders and aggressive conduct. (Apsche, 2010). Adolescents with similar psychopathology are unable to override the automatic reactions that come with their emotional regulation. Therefore, at the heart of MDT functioning is the core beliefs being evaluated and reconstructed through Aaron's (Beck's 1996) theoretical method (Appsche, 2010), location modes are effective sub-organizations of an individual's personality. According to Beck, modes are composed of four interconnected networks, which include (a) cognitive (b) afferent (c) motivational and (d) behavioral elements (Apsche, 2010; Beck; 1996). It was originally developed as a proactive strategy in response to trauma and abuse (Beck, 1996). The primary goal of this

remediation method is to work closely with youth to explore how their modern-day malicious conduct from being socially unacceptable to appropriate with regular healthy responses to trauma and stress. (Bayles, Blossom, &) Upsche, 2014). Thus, MDT has then proved to be very good in remission of physical and sexual aggression among aggressive male youth (Apsche, 2010).

To understand MDT it is necessary to know what a mode is and how it is used in MDT methodology. As stated earlier, modes in MDT consist of four integrated networks that respond actively to stressful and abusive experiences with (a) cognitive, (b) affectionate, (c) motivational, and (d) behavioral aspects. (Apache, 2010, Apache, Ward, & Evil, 2003). Adolescents who best benefit from this measure are those who are perceived as emotionally charged, are treated with malicious behavior, and are "flown off the handle" shortly in response to perceived threats (angry). MDT uses mindfulness techniques to aid in enhancing self-regulation. Mindfulness in MDT is described through a person who is fully aware and accepting of himself as if they are in a moment without judgment (Upsche, 2010, Bayles, Blossom &Upsche, 2014).

MDT helps patients learn that emotion flow can be better utilized by other high-quality methods, just like DBT. This remedy focuses on these strong feelings of fear, anger, and anxiety and teaches children how to overcome their poor urges and attachment (Apsche, Ward, & Evil, 2003). The use of MDT is seen as a good therapy with this population.

Dr. Upsche has observed that traditional CBT can be very decisive and intrusive on adolescent emotions. While working with these children with extreme and aggressive behavior problems, they found that CBT brought re-development to them which was done with MDT, and intensified their resistance to therapy (Apps, 2010). But, through MDT's software, the young are confident with the therapist and have opened up specific and profitable avenues of coping. This can be solidified using a special assumption for MDT methodology known as "recognition, clarification and redirection" (VCR) (Apps, 2010). VCRs are used to provide tools and preparation for adolescents to gain their own unconditional acceptance and validate their journey to know (Apsche, 2010).

Therefore, unlike CBT, MDT accepts adolescent beliefs as truth, no matter how irrational it may be to others (Bayles, Blossom, & Apps, 2014). This helps to create the necessary strong bond between the therapist and the adolescent. It also helps to confirm what the adolescent believes to be true and what is appropriately true and helps them see the difference between what is real and what is true. In a way, it raises the dark cloud and provides a smooth window to the reality of their modern fact (Jennings, Apsche, Blossom, &Biels, 2013). Doing this on all MDTs helps teens to find out who they are explicitly and how to accept them in the context of their cutting-edge environment.

Comparison between MDT, DBT, and CBT

On the surface, it may additionally seem that there is not much difference between MDT and DBT. Both of these treatments have been tested as amazing approaches to increase high-quality social exchanges in these multi-problem adolescents. However, one distinction is that MDT no longer uses CBT theories to the same extent, as the underlying controversy of their beliefs, thoughts, and feelings in the CBT approach causes childhood to reclaim the progress that MDT and DBT have (Murphy and Shiva, 2012).

On the evaluation of the medical effects of MDT versus CBT, MDT has been shown to provide the most desirable outcome in dealing with aggressive and multi-problem adolescents and is significantly more OK in meeting medical goals. Thus, MDT has been developed as a spinoff of CBT to help discover a reliable approach that is more appropriate for treating youth with more than one problem managing their general resistance to treatment. (Upsche, 2010; Epsche, Bass, & DiMeo; 2011). Furthermore, when MDT is the opposite of CBT, it is more effective for physically severe and life-interfering behaviors such as physical aggression and self-harm (Apsche, 2010; Selfish and Apsche, 2014).

Both ACT and DBT use Mindfulness as a way of thinking to facilitate change. Although the initial concept of MDT no longer incorporated explicit mindfulness techniques, it was added in the early 2000s to a customized shape that was appropriate for MDT's adolescent target population. As such, workouts are short, simple, and focused on breathing, visual concentration, mindful walking, and guided imagery that is concise, effective, and within the competence of adolescents (Apps and Jennings, 2013), while regular person mindfulness apply for there is a demand for prohibitive time in training (Carmody & Bayer, 2009). In addition, DBT has incorporated some aspects of

CBT's training capabilities for adolescents to change their behavior and learn new and more efficient behaviors.

By using these treatment options, adolescents have been given the opportunity of early life to exchange their slanderous behaviors and bad coping mechanisms, acknowledging the fact, and now their belief tools (apps, bass, and Backend), can discourage (2012) and certainly change (2014).

In addition, both ACT and MDT address adolescents' avoidance of difficult and painful thoughts and feelings (Apsche, 2010).

This is accomplished through the use of each cognitive and emotional defect, usually through a process where children teach themselves how to stay away from painful stimuli or triggers. The use of mindfulness has deep roots in both ACT and MDT. As previously mentioned, the mindfulness strategies of MDT are specifically designed for use with adolescents. MDT and ACT additionally use one's own acceptance of themselves as they are in the moment, and moving forward with these unique feelings or moving and accepting them as they leave decisions or attachment (Apsche, 2010; Bayles, Blossom, & apache, 2014). During ACT, MDT, and DBT the principle of validation is used to induce euphoric behavior changes through allowing individuals to accept non-judgments of their emotions. These three treatments have been validated to demonstrate the growing trade in adolescents. Although each of these treatment options uses selected elements of CBT, each applies an amazing focal point to cause this change, encourages a strong physician-patient bond, and measures resistance and drop-outs. Limits mostly, with the exception of common CBT, 1/3 wave cures share ideas of peace of mind and acceptance, and although basic philosophies, theories, and desires may additionally overlap, there are subtle differences that make up their respective theories.

Although there is a rapidly growing evidence base for experiencing third-wave therapy processes, there is no conclusive support for an established and unconditional assumption that any or all of the 0.33 wave treatment schemes are most useful for classical cognitive behavioral therapy. However, particularly therapeutic outcome and attractiveness citations are different among unique affected person populations (Kahl, Winter, &Schweiger, 2012). Research of 0.33 wave treatment options has no longer reached the point of maturity where talk or mediation analysis are able to define - quantitatively or qualitatively - the effect of mediating or moderating the outcome of one-of-a-kind therapeutic elements. Nevertheless, there are subtle, and periodically distinct, differences between specific CBT and treatment plans such as the ACT, DBT, and MDT.

The most excellent and meaningful of these are briefly mentioned in an attempt to illustrate precise treatments in brief detail. It is important to reevaluate that some processes that are routinely conserved with the third wave are not understood as such through their developers. One such therapy is dialectical behavior therapy (DBT). The developer, Marsha Lalhan, considers DBT to be an extension of CBT that integrates acceptance strategies alternatively to a true theoretical deviation (Hoffman, Sawyer, & Fang, 2010). In fact, CBT and other 1/3 wave treatments can be seen as a home of interventions that are based on the thinking that "modifying malicious behavior can reduce emotional distress and not easy behavior" (p. 702). However, there are important philosophical and procedural differences between them.

Theoretical Roots: Both CBT and DBT are based on the fundamental basis of work-cause between cognition, behavior, and emotion. According to Hess, the ACT is no longer an extension of the CBT model but is an improvement of Skinner's radical behaviorism that is reacting with reinforcement according to Operational Conditioning Principles, which are involved with acceptance and mindfulness. The developer of MDT, Drs. Jack Epsche has purposefully moved from acceptance and arbitrariness collectively into the body of Beck's cognitive theory with the help of psychoanalytic object family members and the adoption of the factors of Piagetian schema development.

Resistance to treatment, dropout, and attraction: Despite claims that more moderate 1/3 wave treatment outperform CBT, various factors have to be taken into consideration. First, growing rapidly, the evidence base for the ACT, DBT and MDT are very small. Secondly, the development of 1/3 wave treatments was specifically based on the target population and cognitive content is unique to their perturbations that have been considered CBT ineffective. Nevertheless, third wave treatment plans are reporting very good success with their target population in terms of resistance to treatment - a persistent hassle at many hard-to-treat companies and through the Approval Methodology for Third Wave Therapy Counts —

Dropout, and Attrition (Kahl, Winter, & Schweiger, 2012; Jennings, Apsche, Blossom, & Bayles, 2013).

Sustainability and Relief: Similar barriers related to the previously discussed small evidence-based practice. Here, 1/3 trouble is the relative state of third-wave medical concept and practice, which began only in the late eighties and 1990s. However, there are warning signs that these treatment additions and additions to their respective businesses with an appreciation of the durability of the relapse rate (Bach & Hayes, 2002; Linehan, Comtois, Murray, Brown, Gallop, Heard, &Lindenboim, 2009). Although some effect to the contrary was additionally stated (Forman, Shaw, Goater, Herbert, Park, & Yuen, 2012). Specifically related to children with extreme multi-problem behaviors, MDT research has determined a tremendous improvement in behavioral treatment outcomes that clearly outperformed CBT-based TAO control groups, and at least 18 months (Apse). Bass, and Siv, 2006a; Apache, Bass, &Siv, 2006b).

Evidence base: As mentioned earlier, the evidence base of the ACT, DBT, and MDT is a very small one as opposed to CBT. While CBT is discovered in a hundred thousand peer-reviewed publications, ACT, DBT and MDT are notably lagging behind in this regard (see Table 1). However, the research pool of these new third wave therapies is growing at an exponential pace.

Past Orientation: With the exception of MDT, CBT and 1/3 wave therapy only focus on the current second and try to change behavior in a conflicted or otherwise manner by addressing related thoughts and feelings in real-time. MDT has focused a kind of attention on exploring past experiences with complex psychological activities in the present. The premise is that by incorporating a psychoanalytic aspect in the context of the prevailing situation, the extended notion will bring about an additional spectacular and long-lasting alternative effect.

Controversy and acceptance: In CBT, useless thoughts, feelings, and conduct are disputed as "bad" and there are attempts to change their contents or deny their existence altogether. Newer treatments with AER, DBT, and MDT have instead incorporated the idea of radical acceptance, where psychological activities are the norm and no longer an attachment and discourage identification with it. Instead, a decisive perspective of thoughts and feelings is promoted, which are "de-linked" to the idea of self or reality.

Mindfulness: Mindfulness is described as "an awareness arising through inexperience with the other, of paying interest on purpose in the present day, and non-incidentally." (Hoffmann, Sawyer, & Fang, 2010, p. 703)). In simple experience, CBT no longer uses mindfulness techniques and - in addition to mindfulness-based interventions - focuses on a reflex rather than a reactive response to annoying situations and negative emotions. The presumption is that energetic awareness suppresses the brain's default mode community comfortably, spreading rumors of bad thinking that emerge as self-cycles. ACT and DBT base the methods of mindfulness on standard Buddhist meditation practices, while MDT favors fundamental mindfulness exercises that are easy and fast for children to learn, and non-threatening exploration - "Many Pathways to Mindfulness" "(Jennings and Apps, in press)).

Procedural: Some procedural differences between CBT and 1/3 wave therapy are obvious. In general, CBT has a tendency to be more structured versus an ecological mostly procedural method of 0.33 wave treatments. It is interesting to see that this has also been raised as a challenge in the context of theoreticians and therapists to "drift" from mounted frameworks and protocols to reduce fee functioning and integrity. Of these, MDT is considered to be the most procedurally strict (Swart & Apes, 2014), with a structured and sequential evaluation and medical procedure.

Behavioral Strategies: Where common CBT focuses on changing or modifying the psychological opportunities that people travel to achieve the desired behavioral outcome, changing the function of these events as an alternative focus of attention 1/3 wave treatment.

Furthermore, CBT emphasizes experiential avoidance techniques rather than acceptance as ACT and MDT do. As such, CBT tries to "identify and refute maladaptive cognitions with the goal to exchange the emotional response associated with them" (Hofmann, Sawyer, & Fang, 2010, p. 706), whereas ACT and MDT creates and recognise an acceptance of problematic thoughts, feelings, and behavior as they occur. The ACT is therefore response-focused in contrast to CBT that is antecedent-focused. MDT seems to be somewhere between these two, with a major focus on existing events, however a secondary focal point on underlying previous experiences to become aware of and anticipate triggers that can help with emotion regulation.

Techniques: Although there does no longer seem to be primary variations in the ideas and dreams of CBT, DBT, MDT, and ACT, (Brown, Gaudiano, and Miller 2011) contend that variations are more often than not evident in the real strategies that are applied. We have already inferred that 0.33 wave therapies make use of extra mindfulness and acceptance techniques, but Brown and her colleagues additionally pronounced an increased mentioned use of existential-humanistic strategies to tackle the challenges of everyday existence in a balanced and holistic way, although the experiences and influences of the previous are given a good deal much less emphasis than those in the present. Functional contextualism has distinctive importance; strategies are functionally described rather than topographically special as from the place it originates. Exposure techniques are greater commonly used in 1/3 wave therapies, which illustrates the emphasis on ordinary conduct strategies in lieu of cognitive restructuring techniques, although the latter is prominently employed in the "Validation, Clarification, and Redirection" (VCR) step that is a unique process aspect in MDT. Hereby the consumer will become aware of dysfunctional cognitions and their underlying beliefs, whilst growing and attempting out purposeful preferences "one small step at a time", on a continuum.

Therefore, in general, there are possible extra similarities than differences between and among CBT and third wave therapies than typically assumed, though the delicate variations are purposely directed at goal populations and conditions, which amplify their purposeful effect. The ideas of mindfulness and acceptance are frequent to this wave therapies, whilst the previous is, for the most part, discounted, with the exception of MDT, in which past experiences are explored in relation to the journey of their remnants in current psychological events.

The future of 0.33 wave therapies, in precise MDT
The concerns that we have temporarily referred to earlier than want to be greater entirely examined, with specific emphasis on the impact that it has on treatment effectiveness, reliability, and generalizability. Of specific interest is the validity of the underlying scientific and theoretical framework, as nicely as how each remedy approach embodies the standards conceptually and in proper practice. It would also be extremely rewarding to apprehend and differentiate the mediating and moderating influence of each procedure factor on the trade impact and therapeutic outcomes. Thereby, we can come to quantitatively admire to what extent and how method factors such as mindfulness techniques contribute to change.

Such studies have already been performed for CBT, however, are nevertheless lacking in the third wave therapies. It will additionally make contributions to a perception of how and why CBT reportedly fails for sure populations, whilst 1/3 wave treatments seem more wonderful to a lesser or increased extent. We presume, with excellent reason, that each emotional sickness is characterized through cognitive content material that is precise to that disorder, but to what diploma can protocols be tailor-made and/or generalized to improve therapy for the same, other, and broader populations?

The builders of ACT (Stephen Hayes), DBT (Marsha Linehan), and MDT (Jack Apache) have already accomplished a sterling job to promote and establish the evidence base of their respective approaches. Now, persevered effort is required to better understand and recognize the variations in mechanisms and applications of these respective therapies. We do not necessarily need more rising themes, however, perhaps a unified and bendy utility as an alternative that remains in the realm of scientific concepts and validity.

Chapter 3.

The Unique Connection Between Mindfulness and Emotion Regulation.

Heartbreak mourning is recognized as a severe psychosocial stressor that can trigger a variety of intellectual and physical disorders, and long-lasting unresolved grief has a dangerous impact on intelligence functionality. Literature has well documented Mindfulness-Cognitive Therapy (MBCT) as an environmentally friendly measure to improve a wide variety of populations, particularly related to mood and cognition. However, little interest has been devoted to the nervous system in relation to the cognition of bereaved individuals following MBCT intervention. In this study, we recruited 23 bereaved participants who had lost an older relative within 6 months to 4 years to attend an 8-week MBCT course. We used self-reporting questionnaires to measure emotion constitution and functional magnetic resonance imaging (fMRI) with emotional stroke challenge among bereaved participants to evaluate MBCT effects on government participants.

Self-reported questionnaires showed elevations in mood fluctuations and grief remission, difficulty in emotion regulation, anxiety, and depression following MBCT intervention. The fMRI evaluation demonstrated two scenarios: (1) the activity of the frontal-parietal network barely declined with significant elevations in response times to inconsistent trials; (2) things to do in the posterior cingulate cortex and thalamus were positively correlated with the Texas Revised Inventory of Gory, influencing emotional interference on cognitive functions. The results indicated that MBCT facilitated government control by reducing emotional interference on cognitive functions and advised that the 8-week MBCT intervention appreciated both executive control and emotion law in bereaved individuals.

The death of a nurtured victim is identified as a severe psychological stressor, resulting in a time of extreme danger of intellectual and physical distress (Strobe et al., 2007; Buckley et al., 2010). The emotional response to bereavement, commonly known as grief, involves various biological, psychological, and behavioral symptoms. Mourning is related to the signs and symptoms of bereavement depression, distress, and anxiety (Byron and Raphael, 1997; Sasuke et al., 1997; Taylor et al, 1999). These poor effects, therefore, induce deprivation in difficult individuals with health issues (Beem et al., 2000; Buckley et al, 2010; Assareh et al., 2015) and undoubtedly cognitive functions (Clayton et al, 1971 ; Gundel et al., 2003; Rosnick et al, 2010). In addition, large evidence has shown increasing mortality and morbidity prices in the early months of mourning (Young et al., 1963; Lichtenstein et al., 1998; Christakis and Ivasana, 2003; Hartal, 2007). Though the widows have probably identified as a risk (Caprio et al., 1987; Buckley et al., 2010).

Mindfulness is routinely described as paying attention to inside and outside experiences in the present moment without judgment or reaction (Kabat-Zein, 1994). To foster our ability to focus on the current second and egocentric observers of internal cognition, which are adopted during mindfulness training, to promote cognitive attribution and to respond to emotional or cognitive triggers and to decorate adaptive techniques whatever the assumption is (Bishop, 2004; Kang et al., 2013). Neuroimaging research of mindfulness additionally proved that mindfulness can reduce the pastime of the amygdala and increase the thickness of the cerebral cortex (Hölzel et al., 2011; Kral et al., 2018). Many key factors such as attention regulation, body awareness, emotion regulation, acceptance, self-transcendence, and cognitive flexibility are properly developed with the help of meditation training (Shapiro et al., 2006; Moore and Maliniski, 2009). is. Et al., 2011; Wago and Silberswig, 2012). Research on mindfulness has been documented to be truly useful for people's emotional regulation (Hölzel et al., 2011; Teper and Inzlicht, 2013; Roemer et al., 2015), such as Major Depressive Disorder in Psychiatry (Ma and Teasdale) for. (2004) anxiety Disorder (Goldin and Gross, 2010). Mindfulness-based interventions were found to reduce relapse or recurrence of depression (Teasdale et al., 2000) with similar effectiveness to antidepressant treatment (Kuiken et al., 2015) to promote the ability to control emotion such as administration of signs and symptoms of anxiety (Hoge et al, 2013) and stress (Shapiro et al, 2005). Remarkably, mindfulness interventions have been purported to enhance government facilities (Taper &Inzlicht, 2013), such as working remittances and continued interest and interest switching (Chambers et al, 2007; Jha et al, 2007; Zeidan et al. 2010)). Furthermore, based on an observation paper performed by Gallant (2016), the benefit of Mindfulness over Prohibition is consistently recognized compared to other governmental functions.

Mindfulness, in view of the reviewed literature on truly supportive outcomes of mindfulness-based cognitive therapy (MBCT), has thus been developed as a group-based intervention that teaches participants (1) the content of cognition or psychiatric experiences, having awareness in addition to fusion with; (2) staring emotional stimuli corresponding to worrying events, such as sad and poor perception, especially from the loss of one's life; (3) accepting the taboo of emotion and then switching to interest for an unbiased object (e.g., physical sensation or breath) (Teasdale et al., 2000; Segal et al., 2002, 2013).

MBCT has been demonstrated to be a formidable success in helping sufferers with depressive and bipolar disorders for stressful regulation, and comprehensive attention and inhibitory control (Kuyken et al., 2015; Lovaas and Shuman-Olivier, 2018). Although MBCT has an amazing effect on bereaved individuals, the ultimate problems are whether emotional dissatisfaction enhances cognitive functions and what the underlying neurophysiological basis is. Next, we examined MBCT modifications on bereaved people using a sequence of self-reported questionnaires with numerical stroke-work with the arithmetic-FRRI experiment, and self-reported questionnaires of emotion constitutive abilities. We, therefore, hypothesized that cognitive function would be expanded in bereaved members after MBCT training, with similar individuals evaluated before and after the intervention of MBC.

Materials and methods

We used experimental sketches with self-reported questionnaires and fMRI sessions before 8-week MBCT coaching (pre) and after MBCT intervention (post). A within-subject sketch was used to evaluate cognitive performance and fMRI activation adjustments associated with the 8-week MBCT intervention.

The participants
Twenty-three bereaved subjects (21 women and 2 men) of age between 25 and sixty-six (mean = 48.35, SD = 11.14) within 6 months to 4 years of having at least one elder relative wrongfully and self-reportedly participated in unresolved misery.

Study
Contributors were recruited by everyone through advertising and marketing at the National Taipei University through the city of Taipei and in Internet forums of education. All participants were native Mandarin speakers. Participants received a free 8-week MBCT course. The exclusion criteria were: preoperative Mindfulness Meditation experience, history of psychiatric disease, use of prescription drugs and inability to MR. Four of these were excluded for the following reasons: two moved out of the city before the intervention of MBCT; dropped out in the middle of an intervention; one could no longer perform MRI scans after the intervention. The complete data units were accessible to participants for 19 and 20 fMRI scanning and questionnaire analysis as a result, respectively. Written knowledge was obtained from each participant according to authorized suggestions and study

protocols through the Office of Research Ethics of the National Taiwan University.

Self-reported questionnaire
The severity of grief is assessed through the Texas Revised Inventory of Gord (TRIG) (Fashingbauer et al., 1987), which is composed of components of past behaviors and current emotions. As the purpose of this study is to evaluate the results of the MBCT intervention on current cognitive and emotional responses, only the grief portion of the stock was used in the current study. In examining the tendency of anxiety levels and severity of depressive symptoms, Generalized Anxiety Disorder-7 (GAD-7) (Spitzer et al, 2006) and the 18-item Taiwan Depression Scale (Leila al., 2000) were adopted respectively. In order to consider the diploma of the emotion regulation problem, difficulties included employing the Emotion Regulation Scale (DEES) (Gratz and Roemer, 2004).

The Five Facet Mindfulness Questionnaire (FFMQ) 39-item questionnaire has been used extensively to sensitively assess cultivated traits using pique (Baer et al, 2008). Since the validity and reliability of the Taiwanese model of FFMQ have been confirmed, we employed another location (Huang et al., 2015) once in the T-FFMQ and suggested a complete rating in this study.

Mindfulness-based cognitive therapy (MBCT)
The intervention followed a group-based MBCT application (Segal et al., 2013), including weekly conferences once a week for 8 weeks (with a length of 2.5 hours) and every-day home practice (30–40 minutes a day), of course. During the crew sessions, participants were informed of (1)

guided meditation, (2) experiential practice, and (3) participants' every-day practices with skill coaching and in-class practice led by a coaching therapist. The crew doctor, the first author of this study, is a certified grief therapist and has more than 3,200 experiences to facilitate group intervention. Specialized in-class guided meditation protected rejuvenation scans, seated meditation, compassion meditation, and yoga. In addition to group sessions, participants were motivated to perform mindfulness exercises during the day to obtain help using trendy audio-recordings and to record examples of each of their day-to-day exercises at home, which were evaluated as weekly root sessions. In addition, an additional 2-h introduction session of "psycho-physiological responses to loss and acknowledgment of psychological thoughts for loss" specifically designed for bereaved humans was brought before the popular MBCT program.

Practical work

We performed numerical stroke duties in assessing government management function for bereaved participants in contemporary participants. Two types of magnitude decisions were included in this task: the physical shape project and a numerical magnitude task. In the physical dimension task, members are presented with a pair of digits and were urged to decide which digit was physically larger, ignoring the numerical magnitude of the digit. On the other hand, is a function of numerical magnitude, contributors were looking at a pair of digits and were asked to indicate that they were numerically larger, ignoring their physical size.

In both tasks, character digits between 1 and 9 except 5 have been used to form pairs of digits, and pairs of digits in an Arial font with two particular font sizes (55 and 73) for the physical measurements of the item has been introduced to manipulate. For each session in the numerical stroke task, the 4 segments were designed with two blocks of congratulatory position and two blocks of inconsistent position, and the presentation sequence was randomized to minimize fatigue effects. Each block began with a 30-s fixation-cross resting period and a 36-s presentation of digit-pair trials, with each trial consisting of 1-s fixation and 1-s presentation of stimuli. Members were requested to make decisions with the help of button presses within the duration of 1-s stimulus presentations. In analog blocks, the digit that was once larger in magnitude was also larger in physical size. In unadjusted tests, points that were larger in magnitude were smaller in physical size. The total acquisition time for the numerical stroke challenge was 264 s.

MRI data acquisition
The MRI experiments were performed on a 3T PRISMA scanner (Siemens, Erlangen, Germany) at the National Taiwan University. All visual stimuli, delivered with the aid of a projector with E-Prime software, are reflected through the mirror settings.

To reduce the artifacts of motion due to speaking, the main role of the members was once stabilized by the use of thermoset plastics. The scanning protocol employed a high-resolution T1-weighted anatomical image using the 3D-MPRAGE sequence and two objective classes of numerical functions using single-shot, gradient-echo-based echo-planar imaging (GE-EPI).

Sequence

The target parameters for the 3D-MPRAGE sequence were listed below: 192 × 192 × 176 matrix size; 1 mm × 1 mm × 1 mm in-plane resolution; 900 ms inversion time; Repetition time (TR) = 1,900 ms, echo time (TE) = 2.28 ms; Flip perspective (FA) = 9 °; Bandwidth = 200 Hz / pixel; NEX = 1. The total scan time is 5 minutes 21 s. Subsequently, the practical periods shared the same geometry settings: 37 axial slices (FOV = 220 × 220 mm2, sixty-four × sixty four-plane matrix size, and 3.4 mm thickness) obtained in an interleaved manner, aligned with the anterior commissure of the entire brain.

Post-fissure (AC-PC) line after coverage

The GE-EPI scan protocol imaging parameters used were as follows: TR = 2 s, TE = 35 ms, FA = 84 °, bandwidth = 2368 Hz / pixel and the total acquisition time for each session was 264 s.

Data analysis

The processing of all MRI facts has been analyzed by the Analytical Functional Neuroimaging (AFNI) software package deal (Cox, 1996) and the FMRIB Software Library (FSL) (Jenkinson et al., 2012). Montreal Neurological Institute (MNI) template space, 6-mm FWHM spatial smoothing preprocessed task-fMRI statistics set including slice-timing correction, action correction, spatial normalization. After preprocessing, one-by-one congratulations were determined by a model by determining a canonical hemodynamic response function (HRF) with the mission paradigm and task-fMRI for inconsistent stipend as the beta size was regained. The graph matrix in the first-level fixed-effects assessment consists of two registers of main interest: one for interference (pre-versus-MBCT) and another for circumstantial (analogous or inconsistent) contrasts. In addition, six additional registers prescribe head action in preprocessing as covariates without interest.

Parameter estimates from incoming contrast maps were then entered into a second-level random-effects analysis to understand intelligence areas that were largely active with the help of a difference in participants. Vowel-wise one-sample t-tests were performed to become aware of active vowels associated with mission prerequisites both before and after the MBCT intervention. As an improvement for multiple comparisons, a significance level of corrected p <0.01 was determined in conjunction with an unknown threshold of p <0.001 and an individual cluster measure of 90 contiguous tones. To explain the 8-week MBCT effect, we performed a vowel-based paired pattern t-test for each challenge condition. Considering the relaxation of statistical sensitivity on the interference effect due to the noise amplification of the substation method, more than one comparison was conducted once with a coping talent mask, and using a standard significance step of p <0.05 of the uncontrolled rage. The mixture was performed of p <0.005 and the individual cluster dimensions of 103 contiguous vowels. Group-level activation regions have been served as pre-known for the following field-of-interest (ROI) analysis.

To keep away from the double-dipping problem, we extracted ROI values based on an automated physical labeling (AAL) template (Tzourio-Mazoyer et al., 2002) and assessed brain-behavior correlation with questionnaire scores. Furthermore, for the reason that the thalamus is generally concerned in emotion regulation (Grecis et al., 2007; Peng et al, 2012), the thalamic ROI was hypothetically selected for correlation assessment in this study.

The leading behavior result was calculated to extend into RT, over and after the common RT for stimuli (and (conformity) / elicitation × × 100) for stimulating stimuli (Colcombe et al., 2005). Stimulating stimuli increase RT. The percentage increment measurement was once derived from mirror interference using the difference in base RT. Only correct responses were blanketed in the effect measure. Paired t-tests were commonly performed between MBCT interventions. However, if the information distribution violated the general assumption using the Shapiro – Wilk test, a non-parametric Mann – Whitney test was used. In addition, two-way repeated-measurement analysis (ANOVA) was performed on within-subject elements (pre-MBCT, post-MBCT) and conditions (congratulations, inconsistent) over time.

Results of self-reported questionnaires

To have a look at the self-reported questionnaire regarding the effectiveness of the MBCT intervention, participants' post-MBCT scores contrasted with their pre-MBCT ratings on FFMQ, TRIG, DEES, depression, and anxiety ratings, using repeat use. Are through Of the t-test. Mindfulness and all ratings of psychological variables were large at p <0.01. For example, the effect of TRIG was substantial at t

(19) = − 3.98, p <0.001, d = − 0.89. Similarly, t-FFMQ was found to be significant at t (19) = 3.57, p <0.01, d = 0.80. The MBCT intervention was associated with effect sizes (Cohen's D) of − 0.89, − 0.65, − 1.17 and − 0.76, respectively, to reduce the issue of grief, anxiety, frustration, and emotion regulation, respectively, whereas the effect in the improvement in size was once 0.80. Mindfulness Level. These strong effect sizes indicated that MBCT significantly reduced deviant feelings, and expanded the phase of thoughtfulness among bereaved participants.

This finding indicated that the Mindfulness Intervention has high-quality outcomes on bereavement regulation, bereavement alleviation, and makes the Mindfulness Score larger because broader variations were found for ranking comparisons between post-MBCT and pre-MBCT. In addition, to take a look at the relationship between Mindfulness and Efficiency Reactivity, correlation analysis was performed between the FFMQ scale and the post-MBCT intervention deviant emotion state, including TRIG-Present, GAD-7, Depression, Durso Scale. Results indicated that the submitted mindfulness state was particularly negatively correlated with all poor emotion states, TRIG-Present r = -0.52, p <0.05; GAD-7R = -0.70, p <0.001; Depression r = -0.59, p <0.01; DEES r = -0.91, p <0.001.

Behavioral consequences

The accuracy and RT of all participants were recorded while they performed the Stroop mission in the scanner. Accuracy did not hold substantial current differences in MBCT using the Mann – Whitney test, ns (pre-MBCT accuracy = 86.2% and post-MBCT accuracy = 86.8%). We found that to compliment RT (both in numerical magnitude and physical size) was no longer reliably distinct between pre- and post-MBCT, t (37) = 0.59, ns, whereas in RTs for inconsistent trials, t (... 37) = -2.4, p <0.05. Significant interplay effect of RT was explored in Intervention × Status, F (1,18) = 6.73, p <0.018. In addition, an additional assessment of proportional intervention scores, unbiased by differences in base RT, showed a wide decrease after MBCT intervention (13.1 to 8.7%), p = 0.002.

Neuroimage Results on Numerical Stroop Tasks

Figure 1 demonstrates the substantial recruitment of the dorsal attention network (DAN) to a session of numerical stroke functioning, and the detailed facts of brain regions were proved in Tables 3 –5. The degree of DAN activation was once moderately reduced after MBCT, while maintaining accuracy, but the results of inconsistent tests involving a greater cognitive-weight of inhibition, anterior cingulate cortex (ACC) and posting cingulate revealed vast inactivation in both cortex (PCC). Similar to the RT results, congruent tests did not produce current MBCT results in brain recruitments, whereas brain recruitment in incompatible trials after MBCT confirmed a drastic decrease in PCC/precuneus corrected under alpha PCS p <0.05. The interplay effect of the fMRI results (Intervention × Condition) showed single impaired activation located on the PCC, similar to the inconsistency result. We further evaluated associations related to anomalous conditions of the numerical Stroop task between neurological evaluation and field effort intelligence (PCC and hypothetical thalamus).

Figure 2 shows the heavy correlation (A) between TRIG and PCC; (B) between anxiety and PCC; And (C) TRIG and thalamus, Spearman's $\rho > 0.33$, $p < 0.05$. Figure 2 shows that very little grief and nervousness was related to reduced neurological activities of the PCC or thalamus that were related to numerical stroke work.

Discussion

The reason for the contemporary finding was to examine the sentiment legislation in mourning and the facilitative impact of the 8-week MBCT on government work. As all feelings, thoughts, and body sensations are taught to be accepted with non-judgmental acceptance, bereaved participants abstain from the horrific stories of these sensations associated with them (Kabat-zinn, 1994). As anticipated, bereaved contributors after the MBCT intervention suggested drastic reductions in grief, anxiety, depression, and difficulties in emotion regulation, as well as increases in the state of mindfulness. These spectacular changes can also stem from the fact that bereaved members have fostered a monitoring capacity on their grief responses, with emotional acceptance, a non-judgmental attitude, and switching interest again for the present. Over time such practices can sharpen participants' emotion skills in their everyday lives, allowing them to relax physically. From the group discussion, many bereaved stated that they slept better and felt greater vitality in each of their day-to-day lives after MBC intervention.

Our first intention is to find out about cognitive improvement after bereavement mitigation after MBCT intervention. The Stroop task is a cognitive measure to examine government control function, which allows humans to overcome impulses and override automatic behavior. In this work, we have replaced the simple color-word Stroop project for bereaved individuals, considering the inability to pay attention to the high cognitive load in the established section (Huang et al, 2012). Following Mindfulness Training, we observed lower intervention RT scores when performing numerical Stroop enterprise in participants who were in mourning, reflecting multiplication cognitive manipulation performances (Colcomb et al., 2005) associated with degrees of rejection post-MBCT.

Previously, Tepper and Inzlikt (2013) and Tepper et al. (2013) confirmed greater emotional acceptance and improved performance among Mindfulness Mediators in EEG based assignments, suggesting Mindfulness Training such as Meditative Practice provides control to the government. Gallant (2016) additionally supported this announcement in his review article (Gallant, 2016). Furthermore, even though Tepper and Inzlicht (2013) hypothesized that temperamental practices promoted government manipulation before improving emotion-regulation. However, we can no longer confirm the inference based on our observations in bereaved populations. Further experimental designs are warranted to reflect this notion.

Secondly, practicing temperance allows bereaved individuals to abandon highly intuitive tasks and minimize emotional interference with cognitive functions. Therefore, we used the fMRI test to uncover the underlying neurophysiological basis of MBCT-based cognitive enhancement in bereavement. The results confirmed numerical stroke activation of bilateral Dan, including the center frontal gyrus and the most efficient parietal gyrus, comparable to the previous file (Huang et al., 2012).

Overall, following the MBCT intervention, the low DAN pastime implied a lower cognitive load of anxiousness in numerical cognitive function. Meanwhile, in bereaved participants the ACC and PCC diverged from coping with inconsistent trials, indicating a cross-network interaction in the task. Interestingly, default-mode community (DMN) deactivation was hitherto used in coping with the excessive cognitive load of working recall tasks (Liang et al., 2016). We hypothesized that mismatches arise from over-activity of DMs with excessive intrinsic thoughts in a population, interfering with their normal functions in cognitive performance.

Similarly, Gundel et al. (2003) reported talk of PCC and medial frontal gyrus among bereaved women in response to bereaved words, indicating that these areas are concerned with affective processing. Christophe et al. (2016) elaborated in their observations the function of the DMN core in internal and intuitive thoughts, influencing conceptual and emotional processing. In our report, the PCC in the numerical stroke mission had a very high-quality correlation with TRIG and anxiety, suggesting that PCC's strong beta level charges coupled with strong bereavement mismatch numerical function too.

Finally, we no longer detected hypothetical correlations between T-FFMQ and PCC/thalamus activity. One possibility is that DMN-related regions may be more concerned in spontaneous activities of emotional arousal, as an alternative, than participating directly in the mind. A second occasion is that even the mindfulness dose of the traditional 8-week MBCT protocol may be insufficient for relief from emotion discrepancies in bereavement. Further MBCT-based research is warranted to validate the hypothesis on bereaved populations.

Although the current results suggest that MBCT leads to more appropriate emotion regulation in practical subjective assessments and government manipulate fMRI environments, we no longer examined anxious genius activity in emotional arousal. For the concerns of humanity, we try to keep the bereaved participants away from increasing more affectionate excitement. Since we have found that they did not employ direct measurement of emotional sensitivity, it is challenging to say whether the DMN and thalamus were once concerned in lubricant processing. However, the literature gave strong help on the emotional involvement of each PCC, ACC, and thalamus in primary depressive disorder, insomnia, and anxiety disorders (Grecis et al., 2007; Bastian, 2011; Christophe et al. 2016).

Secondly, due to individual differences in participation time, degree of initial frustration, and understanding ability over temperament, the mastering curve to achieve anticipatory mindfulness was diverse. For example, some of them suggested being in a position to consciously relate their positions to the concluding session of the MBCT course, although some suggested that they felt completely high during the interior of the group. Therefore, emotional balance in the same MBCT crew demonstrated strong inter-subject variability, imposing a confusing element on the concluding assessment. Third, Huang et al. (2012) used numerical magnitudes and physical measurements of the numerical stroke to differentiate hemispheric differences between the elderly and the young. However, we were no longer aware that there are subsequent changes to the work currently done.

Also, a realistic project in this addresses the use of treatment as a manipulation group in addition to treatment, due to the fact that none of the bereaved participants admitted to enroll in the management group. Nevertheless, the nomination of energetic control may be circumvented, due to the fact that literature reveals that understanding of grief measured via TRIG after the bereavement factor was overcome by eight weeks of MBCT (Zisook et al) took years to recover again to a regular condition. (1982). Similarly, Tseng et al. (2017) state that it took about four years for the depression level of the Taiwanese mourning stage to be restored. Evidence confirmed that grief has long-term effects of prevention without bereavement and is no longer without problems within eight weeks. Further investigation is done to aid statements on the subject of grief mourning.

In particular, working towards an 8-week Mindfulness Education has helped reduce grief, anxiety, depression, and improved their mindfulness state among bereaved individuals. Mindfulness training no longer benefited from mere emotional regulation, although emotional interference on cognitive functions was also reduced. The results show a decrease in RT ratings in numerical stroke challenge among participants with decreased intervention RT ratings, leading to an increase in government control function. In addition, the positive talents of the PCC and thalamus showed an interfering effect on the Stroop task, but the post-MBC intervention reduced the outcomes of the intervention of the PCC and thalamus, which is associated with bereavement. Based on the truly useful effects of MBCT interventions, we encourage the bereaved population to practice mindfulness education to avoid excessive emotional stimulation and to maintain a satisfactory daily life.

Chapter 4.

Skills to Help You Find Focus in the Present Moment.

You have said this many times before that it is important to stay in the present moment. You have additionally heard the same piece of advice:

"Don't get stuck thinking about the past or the future - live in the now!"
"Stay in your life."
"You have this moment. Do not let it slide."

All these (possibly overused) proverbs boil down to the same basic message: it is important to persist in the present moment.

Our contemporaries live in the twenty-first century, now it is not easy. There is something constantly unfolding that we want to do together or anticipate, and our lives are so well documented that it is not easy to get lost in the past in any way.

Given the fast pace and nerve-racking schedule, most of us have a new level of anxiety, tension, and sadness. You, too, cannot understand it now, although this tendency may be sucked into the past and the future can always spoil you and feel in touch with yourself.

The therapy for this condition is one that has been heard by so many humans about pronouns: conscious awareness and "now." Dedication to living in. Living in the present moment is a solution to a problem that was no longer in your thoughts.

You might be questioning that it all sounds great, but what does it honestly suggest to "live in the present moment"? Besides, how should we live in anything in the present?

Psychology of living in the present
Living in the present is no longer just an arbitrary word or a popular phrase - it is a recognized and evidence-supported lifestyle that psychologists are increasingly supporting for people struggling with anxiety and stress in their day-to-day lives.

What is the meaning of the present moment?
Being in the present moment, or "here and now," is the skill that we are conscious and aware of what is happening at the moment. We are not distracted by using concerns about the past or the future, though set in the here and now. All our interest is focused on the present (Thum, 2008).

Why is Being Present Minded Important?
It is important for the present mind to be healthy and happy. It helps you fight anxiety, reduces your stress and worries, and drives you forward and connected to yourself and the people around you.

Although it has become a well-known subject matter in the latest years, living in the present is not just a fad or cutting-edge lifestyle tip, it is a way of life that is supported through proper science.

Being present and keeping our capacity in mind no longer makes us completely happy, it can also help us deal with pain more effectively, reduce our stress and our health. It may lessen the impact, and increase our ability to withstand terrible feelings like anxiety and anger (Haileywell, 2017).

Why it can be hard to live in now
It is so hard to live in the now because we are usually influenced to think about the future or to think about our past. Advertisements, reminders, notifications, messages, and signals all often lead to past or future.

Think about how often you are busy doing other things, possibly completely absorbed, when suddenly you hear a "Ding!" right here and now.

If you are like me, your response is probably "just about never". Our telephones are high-quality pieces of technical knowledge that allow us to get such a great deal and do it more effectively than ever before. But we honestly want to take a run from our phone at least once.

Other factors that contribute to our inability to live include:

We regularly edit the horrific components of our experiences, making our past exciting.
When we live in the present we face a lot of uncertainty, which can cause anxiety.

We definitely have a tendency to roam!

This can counter these factors hard, but fortunately, we are no longer slaves to dispose of our minds (Tlalka, 2017). It is possible to overcome our greater negative or harmful urges and make better choices.

Balancing the past, present, and future
Sometimes it is good to speculate about the past and the future.

Where will we be if we look no less than our past successes and mistakes and analyze them? Where would we be if we never deliberated for the future or organized ourselves for the time to come?

In both cases, we will not be in a good place.

Spending some time to question the past and the future is unavoidable for an entire existence, but it is unusual that we do not consider enough about the past or the future — usually our problem with great attention to the past (or here even obsessively) concentrates.
One of the objectives of Mindfulness and a major issue in an entire lifestyle is to balance your thoughts of past, present, and future. Thinking too much about any of them can have seriously bad consequences on our lives, although preserving all three in balance will help us be fully happy and healthy people.

It is challenging to say what the specific right consistency is, but when you are less afraid, you will recognize that you are less afraid, ride a lot less stress on an everyday basis, and find yourself currently in your life. Most explore the part.

How to present and live in the moment
For this healthy balance, try to remember these guidelines:

Think of the past in small doses, and make sure that you are focusing on the past for a purpose (for example, to get a good experience, choose where you went wrong, or the past. Understand the key to success).

Think of the future in little portions, and make sure you're focusing on the future in a healthier, less worrying way (e.g., don't spend painful time thinking about the future, think about the future to prepare for it long enough and then walk).
Stay present for the broad majority of your time.

Of course, following these tips is less complicated, although it will become easier with practice!

How to live in the moment of planning for the future
Probably this would seem complicated to determine refined equilibrium, although it is not as complex as it seems.

When we interact in mindfulness or present moment meditation, we are not ignoring or denying the past or future thoughts, we are honestly choosing not to pay attention to them. It is okay to thoroughly know and label our past and future-focused ideas, categorize them, and become aware of their importance.

The essential factor is no longer allowing yourself to be swept away thinking about the past or future. As Andy Pudukombe of Headspace states,

"... we can be in the present when consciously reflecting on events from the past as being caught hostile, distracted and overwhelmed by using the past" (2015).

When we are aware and present, we do not want to be afraid of getting bogged down in nervous thoughts about our past or our future - we can revisit our past and speculate on what to lose ourselves from.

Using present moment awareness to prevent anxiety
Talking about anxiety, the second consciousness is the best way to reduce how much anxiety you currently have.

Follow these six steps to get more and more in the present and get rid of your own excess anxiety:

- ❖ **Think yourself stupid:** Let go and think about your performance.
- ❖ **Savoring to practice:** Stay away from worrying about the future by experiencing the present.
- ❖ **Focus on your breath:** Enable Mindfulness to give you more peace and make your interactions with others easier.
- ❖ **Find your flow:** Make the most of it using music.

- ❖ **Improve your ability to accept:** Instead of denying it or running away, move towards what is bothering you.
- ❖ **Increase your busyness**: Work on the waning moments of imprudence and notice new cases to improve your will (Dixit, 2008).

Using yoga exercises to connect with the present moment

You probably won't be surprised to hear that yoga is a great way to connect with the present and continues even at this time. There are many reasons yoga is helpful for meditation, but in fact, the biggest is the focus of attention on the breath.

This easy workout will take you straight to the present, even pulling with a stubborn thought that is prone to anxieties.

Another issue related to yoga that allows us to promote our existing second belief is the postures and poses we create with our bodies. You may find that the sooner you get into a correct posture, your thoughts become flooded with stressful thoughts ("monkey mind" through Buddhists). As worrying as it is, it can without a doubt be a well-spoken thing - the skill that we are beginning to reduce our stress and achieve a factor in the place where we are definitely mindful (Bielkus, 2012) and can exercise.

A gentle wave of yoga from one function to another is the best possibility to develop the ability to be present. We mimic the changes that we experience as we transition from work to cooking to cleanliness and everything in between.

Five exercises to strengthen present moment awareness

If a respiratory workout sounds helpful, then you might want to try some different workouts to increase your mood and experience present moment awareness. These are some of the exact ways to start these five exercises.

Mindful body scan

This easy exercise is a high-quality way to get yourself into a temperamental nature and get in touch with your body. Doing this in the morning can prepare you for a great start to your day.

While sitting on your mattress (if you try to lie this way, don't fall!), Take some deep, arbitrary breaths. Notice how your breath enters and exits your lungs.

Starting with your toes, focus on one part of your body at a time. Pay interest in the area and be aware of any sensations you are experiencing (Scott, n.d.). After a few moments of focused meditation, move on to the next part of your body (i.e., after your toes, focus on your feet, then the ankles, then the calves, etc.).

This is not only a desirable technique to put you in a conscious state, but it can also help you get the word out when your body is feeling different than normal. You can usually scan your body for a few minutes each morning to prevent damage or illness that you usually do not notice.

You can check out more about mindful body scans and different workout routines.

Write in Journal / "Morning Page"

Another true workout that can help you set the appropriate conscious tone for the day is to write in your journal. A unique model of the practice that the author supports using Julia Cameron is called "Morning Page".

Here's how to use your journal as a step to a more conscious day.

Before dawn, when you want to investigate matters starting from work or college or your long list, take some time to make your journal or pocketbook and make an entry.

You can create a new web page every day and actually alternatively write an awesome lot that you like to write, or you can try Cameron's morning holiday practice:

"Morning pages are three pages long, a stream of cognition writing, the first element in the morning. There is no wrong way to do morning pages - they are not high art. They are not even" writing "anymore. They are just about anything and everything that crosses your mind — and they are just for your eyes. Morning pages, provoke, clarify, relax, cajole, prioritize, and day synchronize the micro hand "(Cameron, Ann. D., As Scott, Ann. D. Stated).

Whether you follow Cameron's tips or not, take just a few minutes to write down any of the insensitive "nonsense" in your head or log any in particular practical desires, cleaning your head can help you start your day in an aware state.

Imagine your daily goals

Visualizing your desires is a terrific way, which not only makes it more likely that you will follow through on your goals, it can also help you be extra conscious on a simple basis.

When you have set your everyday dreams (see #15 - Define the three daily goals on this list if you want help with this piece), take a moment to imagine each one (Scott, Ann . D.).

See yourself a challenge today to complete each goal and to accomplish each objective. Get as much of a good deal element in your visualization as possible, so it feels real and within your reach.

Ensure that every day goal from your list goes to the next intention and repeat until you have visualized all your daily goals.

Practicing the purpose completion scene can not only help you improve your focus and center of attention, but it can also reduce your stress, increase your performance, beautify your preparedness, and provide you with the extra strength or motivation you want to accomplish the whole thing.

Take a mindful nature walk
Taking advantage of the natural splendor, we have another good way to cultivate more and more mindlessly.

When you feel the need to take a walk - be it a quick trip block or a long walking tour, it is easy to walk mindlessly. All you have to do is interact with all your senses and continue to know what is happening around you and within you.

Be intentional with your awareness; Watch your toes hinge on the floor with every step, see everything there, you have to look round, open your ears to all the sounds around you, feel each breath and exhale. , And usually, be aware of what is happening in each moment.

This exercise not only helps you to connect with your real self, but it also helps you to connect with your environment and improve your cognition of beauty, which is just ready to meet. Known benefits of regular walks include these benefits — reduced stress, high coronary heart health, and accelerated mood — and you have a hand exercise!

Conduct an adorable review of your day
With the help of daytime necklaces, it can be convenient to wear let things slip. To help you catch that conscious tone at the end of the day, try this practice.

Towards the end of your day, perhaps before you finish your "must-dos" for the day or just before going to bed, take a few minutes to observe your day (Scott, n.d.).

Return back to the beginning of the day and figure out your mind workout that stopped it completely. Think about how it made you feel.

Think through the rest of your days, making sure to note any particularly conscious moments or memorable events. Take a list of your mood as you go through your daily routine.

If you want the song of your progress to be of high thoughtfulness, then it is a high-quality concept to write all this in a magazine or diary. However, the point is to give yourself another possibility and end your day on the right note.

Mindfulness Practices and Tools to Use Everyday

One of the pleasant tools to sustain oneself in the present times is meditation. Anyone will meditate, although there are some meditation practices that are particularly close to present moment awareness.

To try this meditation, follow these simple steps:

- ❖ Set aside a simple block of time during your day (e.g., the first aspect of 5 minutes in the morning or before you go to bed).
- ❖ Get in a comfortable position - but not too comfortable now! You will not like to fall asleep when you are trying to meditate. Seating upright can be a pleasant posture.
- ❖ Establish an "inner gatekeeper" to control what comes in and continue to stay out of mind. Instruct the gatekeeper to keep away any thoughts of the past or future for the relaxation of your modern practice.
- ❖ Repeat this phrase silently three times by yourself: "The time has come to know about the present. I let go of the past and the future."
- ❖ Turn your attention in the direction of the sounds you hear. Allow them to focus on you and focus only on the contemporary sound that you are

listening to, no longer what you have heard or any sound you can hear even further.

❖ Pay attention to your physical sensations. Your palms resting on the fingers of the chair or on your lap, your feet are folded on the chair or under you, the feel of your clothes on your skin, any pain or muscle aches. Is there any torsion or quivering or any other sensations you may feel?

❖ Turn your attention to the thoughts going through your head. Keep them in mind as they enter your mind, revolve around your consciousness, then exit your mind. Let every idea pass as they go (for example, the label "hurt" or "happy") and maintain your idea for the next concept to arise.

❖ Finally, place the focal point on your breath. Pay attention to your natural way of breathing and take the word as to how your chest gets up and falls with every breath (Henshaw, 2013).

Although Mindfulness Meditation is an incredibly huge catch-all-term for those types of strategies that help you become more aware and more committed in the present moment, there are some exact types of conscious meditation that you can try.

❖ **Basic Meditation**: Focusing on your breath, a word or a mantra and allowing thoughts to come and go in addition to a judgment.

❖ **Physical sensations:** Be alert to physical sensations such as itching, tingling, sore throat, or tickling sensation and stop them from making decisions, then letting them pass.

- ❖ **Sensory:** Being conscious of what you are seeing, hearing, smelling, tasting and touching in addition to judgment, then labeling them and letting the senses pass.
- ❖ **Emotions:** Permission to be present in oneself except to try to judge or neutralize feelings; Working towards naming/labeling emotions and letting them happen and letting them go easily.
- ❖ **Surfing:** They face cravings with the help of accepting a forbidden decision, seeing how you feel as they hit you, and reminding yourself that they will pass (help guide, nd).

It can be difficult to keep up on the undertaking, but it can be challenging, especially when you are surrounded by a path of steady distraction. In today's always-connected world, variations are nothing but a click away. Even in quiet moments, distraction is at your fingertips as you find yourself checking your Facebook or trying to catch that elusive Pokémon.

The ability to focus on the things around you and to make a direct intellectual effort towards it is essential to learning new things, achieving goals, and performing properly in a vast range of situations. Whether you are trying to finish a record at work or compete in a marathon, your ability for a focal point can suggest the difference between success and failure.

It is better to focus on your intellectual center, although it is no longer suggested that it is always faster and easier. If it was once simple, we would all have to focus on a specific athlete. This will take some real effort on your part and you may have to make some fundamental modifications to your everyday habits.

1
Start through assessing your mental focus

If the first set of statements looks different from your style, you probably already have very good attention skills, but you should improve with a little practice.

If you make also discoveries with another set of statements, you would in all probability want to work on your intellectual focus. This may take some time, but training some top habits and being conscious of your distraction can help.

2
Remove distractions

Admit it, you've seen it. While this may seem obvious, humans often underestimate how distracted they are from focusing on the task at hand. Such intrusions can come in as a radio stigma or possibly an unpleasant coworker who constantly leaves you to chat through your cubicle.

Reducing these sources of distraction often seems simple without a doubt. While it can be as simple as turning off the TV or radio, you may find it more challenging to deal with a handicap peer, roommate, spouse or child.

One way to deal with this is to isolate a specific time and region and request to leave on your own for a period of time. Another option is to look for a quiet place when you recognize that you will not be able to work. A library, a private room in your home, or even a quiet coffee preserve can be a top place to try.

Some strategies you can try to reduce or try to overcome such distractions are to ensure that you are well-rested before the mission and to fight anxiety. Use brilliant ideas and imagination. If you find your mind wandering towards distracting thoughts, consciously, at least, bring your attention back to work.

3
Focus on one thing at a time

While multitasking may seem like a great way to get by very quickly, it turns out that humans are terrible at it instead. Doing a few tasks at once can dramatically reduce productivity and make it harder to improve the details that are really important.

Attentive assets are limited, so budgeting them wisely is important.

Think of your meditation as a spotlight. If you shine a light on a precise area, you can see matters very clearly. If you try to lightly reveal the same amount of light in a large dark room, you'll probably only be able to glimpse the outline of the shadow as an option.

Part of improving your mental focal point is to make the most of the resources available to you. Stop multitasking and instead offer your full interest to one element at a time.

4
Live in the moment

When you are worried about the past, the future, or for some other reason to get out of the present, it is difficult to continue being mentally targeted. You may have heard humans speak about the importance of "being present". It is all about removing the deflections, whether they are physically (your mobile phone) or psychological (your concerns) and are completely mentally busy at the present time.

This concept of being present is also excellent for regaining your intellectual attention. Staying here and now, your attention is sharpened and your mental wealth is respected on the small print which actually counts at an accurate point in time.

It may also take some time but definitely one should work on studies to stay in the moment. You cannot change the past and the future is no more, but what you can do these days can avoid repeating past errors and pave a course for an extra profitable future.

5
Practice mindfulness

Mindfulness is a hot topic right now, and for the right reason. Despite the reality that people have practiced forms of mindfulness meditation for many years, many of its health benefits are fully understood these days.

In one study, researchers had HR experts interact in complex multitasking types of simulations that they were engaged in each day at work. These duties were to be executed in 20 minutes and using cell phones, emails, and text messages with records from some sources, answering phones, scheduling meetings, and writing memos.

Some participants received eight weeks of training in the use of Mindfulness Meditation, and the results found that only those who had acquired this coaching confirmed improvements in concentration and focus. Meditation group members were able to stay at work longer, switched between tasks less frequently, and performed tasks more effectively than participants' other companies.

Practicing meditation can be mastered by meditation, but it can also be easier as a fast and effortless deep breathing exercise.

Quick tips to gain focus
Begin by taking several deep breaths, focusing on honesty and every breath. When your mind starts to wander naturally, slowly and unconsciously return your attention to your deep breathing.

While this may seem like a deceptively simple task, you can also know that it is a lot more difficult than it actually appears. Fortunately, this respiratory effort is something that you can do anywhere and anytime. Eventually, you will find that it becomes easy to get distracted by intrusive thoughts and bring your attention back to that place.

6
Try to take a short break
Have you ever tried to pay attention to the same thing for a long time? After a while, your attention begins to damage the center and it becomes more difficult for your intellectual

senses to function. Not entirely so, but in the end, your performance suffers.

Traditional interpretations in psychology have recommended that this is due to a lack of attentive property, but some researchers agree that this is also due to the brain's tendency to exclude sources of static stimuli.

Researchers have discovered that taking a break too early to shift your attention elsewhere can increase intellectual focus.

So the next time you are working for a long time, such as preparing your taxes or reading for exams, be sure to give yourself an occasional intellectual break. Focus your attention on something unrelated to a mission, even if it is for a few moments. These small moments of relief can mean that you are in a position to sharpen your mental focus and keep your overall performance at a higher level when you honestly accomplish it.

7
Keep practicing to strengthen your focus
Focusing on your mental focus is no longer something that will appear overnight. Even professional athletes require a lot of time and practice to strengthen their awareness skills.

One of the first steps is that being distracted is having an impact on your life. If you are struggling to meet your goals and isolating yourself through insignificant details, then the time has come to start charging high fees on your own. By focusing your mental attention, you will find that you are able to focus more on the matters in life and fulfill them which will surely give you success, happiness, and satisfaction.

Chapter 5.

Reduce Impulsive Behavior.

Impulsive behaviors are those that quickly curb control, planning, or consideration of the consequences of that behavior. Impulsive behaviors have a tendency to be immediately associated with good punishment (e.g., freedom from emotional pain). However, in the long term, many bad consequences can also occur, such as great emotional grief or regret.

Common Serious Impulsive Behavior
In considering their behaviors, it may help to handle some of the more common impulsive behaviors with PTSD. Are there any of the ways that you are currently experiencing emotional pain?

- Eating disorders
- Alcohol abuse or tingling (self-healing)
- Drug abuse (prescription or illegal)
- Self-harm
- Suicidal thoughts
- Gambling

Coping strategies
There is a range of easy strategies to prevent impulsive behavior. If you have animosity with impulsive behaviors, try to see one (or all) of the coping strategies below to see if you can get a better handle on complex behaviors.

Distract yourself
The urge to negotiate impulsive behavior can also be very strong and difficult to cope with. However, these urges usually pass fairly quickly. Therefore, if you can distract yourself from experiencing an urge, you may additionally be able to sit with an urge until it passes. Fortunately, there are many healthy distraction techniques that can be useful in driving a strong urge or emotional experience.

Include your senses in grounding techniques, basically a structure of distraction, as long as you can replace impulsive behaviors with healthy behaviors.

Change your impulsive behavior
Even though impulsive behavior can cause long-term problems, at the moment they are serving a purpose. For example, they may additionally help you deal with emotional pain. Therefore, one way to stop impulsive behavior is to search for another, healthy behavior that can also serve the same purpose. Healthy behaviors to change impulse behaviors include:

- Looking for a friend
- Write about your feelings
- Meet your therapy team or a friend in your group
- Try to find a healthy way to get relief from emotional pain which will no longer be a long term bad punishment for you.

Identify negative consequences
We are motivated through the non-permanent consequences of a behavior. That is, we usually repeat behaviors that work well for us at the moment, regardless

of what their long term bad punishments are. Therefore, it may be beneficial to increase your awareness of the long-term terrible consequences of a behavior. One way to do this is with the help of identifying the short and long term pros and cons of the behavior.

Change behavior results

People go on to behave impulsively as they do something amazing in another (for example, removing anxiety or fear). One way to limit the possibility of impulsive behavior is to remove its short-term spectacular effect. The sooner you interact in impulsive behavior, conduct a chain evaluation without delay as to why you were previously engaged in that behavior. In a chain analysis, you try to join all the links between behavior and consequences. Try these steps:

- Identify behavior for change.
- Identify what happened before the behavior you want to change.
- Evaluate your thoughts and feelings at that time.
- Identify what your thoughts and feelings chose you to do.
- Consider the punishment that occurred.

This process will bring you into contact with all the emotions you are trying to get away from the first area and force you to face and face in a second, healthier way. This can be very useful for rewarding yourself when you do not interact in impulsive behavior.

Many words

It can be very difficult to cope with impulsive behavior, but it is possible. Identify some of the impulsive behaviors you want to change, and the next time you insist on interacting in those behaviors, try one of the above coping strategies. With every success, figuring out a healthy approach to coping with PTSD will become less difficult and easier. Some of these techniques may include:

- Learning about your diagnosis
- Seeing a doctor
- Joining a support group
- Practicing deep breathing
- Engaging in self-monitoring.

Chapter 6.

Increase the Sense of Connection to Your True Self.

Do you think anything in your life has lost weight? Do you understand that you are without any hope and no longer recognize why? Is there something missing and you can't determine what it is? Life can sometimes seem so difficult like we are without a map to show us the way. When we become intrinsically careless and do not understand why, the lifestyle should look more than just hard, tough. You're not alone; many people experience this wrong feeling within their lives. What is everyone missing? What is it that makes people less than expected? For most people, it is because they have misjudged their own life connections and it all surrounds them.

Without connection, we are giving up hope, belief or purpose. When we are not connected, we ended the illusion, even without knowing why, because of the fact that we are lost in reality. What is connected with you and with whom are you connecting? First, you are connecting with yourself; many of us separate ourselves from our lot. We lose who we are and build another existence above our real selves. We perceive our own views as hostile to what we really are. We lose connection with our real self, and we cannot help but lose the method because we have lost ourselves.

There is even a deep connection that we sometimes lose, the connection of the whole thing with us. We lose ourselves through distancing ourselves, usually because we are trying to protect ourselves from fear, anxiety, sadness, or any emotion that provokes escapism. We distance ourselves from life, and the connection is lost when we do. We clip the wires that connect us to exist in two very integral ways. We damage our relationship with everything around us, and we break our relationship with our true self. This is why we have a feeling of being lost, alone and confused. The weight we feel is the absence of knowing who we are and how we reconcile with the relaxation of the world. Everything in the lifestyle is so difficult because we are no longer connected to perfection. In a way, we are doing all this by ourselves and every other person.

When we are performing like any other character or constantly protecting ourselves from the world, we cannot seek or find fulfillment. We have distanced ourselves from the world due to the fact of our fears and emotional reactions to our lives. We created someone else to interact with, manage, or pretend with the world as something other than who we are. How do you find yourself and reconnect and how do you let go of your anxiety so that you can connect with the whole thing around you?

First of all, you need to find and launch all that is definitely no longer you; anything that is no longer required to be absent of your true self so that you are fully connected to yourself again. You will lose the feeling of being alone or lost when you do this due to the fact that you are coming home yourself. When you have launched all of your own components that you are no longer, you are ready to reconnect the world. To be your authentic self, you have to give up the desire to separate yourself from existence and become a part of it again. Fear is missing, fear disappears, and any bad emotional reactions fade due to the fact that you are what is most needed and nothing is without fear.

Connecting yourself and the world is a process that enhances your existence and advances your development. It helps you do what you clearly want to do and find the lifestyle you want to live. It no longer only reduces the feeling of displacement that you are also experiencing, although it restores the feeling of being one with everything, one on one with your real self, and one on one with the world, And from there the whole thing starts to feel easy and your existence will not look so confusing now.

Chapter 7.

Extreme Stress or Difficulty.

Stress is a biological response to stressful situations. It causes hormones to be launched in the body, such as cortisol and adrenaline.

These hormones help to formulate action, for example by helping to increase heart and respiratory rates. When this happens, a health practitioner will likely describe a man or woman as being in a state of extreme alertness or excitement.

Several elements can trigger a stress response, including dangerous situations and psychological pressures, such as work deadlines, exams, and carrying events.

The physical consequences of stress usually do not last long. However, some humans find themselves in almost a state of constant vigilance. It is a constant tension.

Some viable causes of continual stress include:

- High-pressure jobs
- Financial difficulties
- Challenging relationship

Chronic stress exerts pressure on the body for an extended period of time. It can give a variety of indications and may increase the risk of developing certain diseases.

Signs and symptoms

Chronic stress affects the entire body. It can have countless physical or psychological symptoms, which can make the foundation of every day more challenging.

The type and severity of symptoms vary greatly from man to woman.

Symptoms and signs of chronic stress may include:

- Irritability, which can be extreme
- Fatigue
- Headache
- Difficulty concentrating, or inability to do so
- Sharp, chaotic thoughts
- Difficulty sleeping
- Digestive problems
- Change in appetite
- Feeling helpless
- Perceived loss of control
- Low self-esteem
- Loss of sexual desire
- Nervousness
- Recurrent infection or disease
- Health outcomes

In the long run, chronic stress can contribute to improvements in a range of physical and intellectual disorders, including:

- A weak immune system
- Venereal disease
- Gastrointestinal Disorders
- Skin irritation

- Respiratory infections
- Autoimmune diseases
- Insomnia
- Burn out
- Depression
- Anxiety disorders
- Traumatic stress disorder, or PTSD
- Schizophrenia

Managing stress

Long-term stress can appear overwhelming, and a person may additionally feel unable to take control of their life.

However, a number of strategies can help reduce stress limitations and improve well-being.

Some methods of managing stress include:

- **Understanding the signs and symptoms:** These warning signs can vary, but if a character can recognize their own signs of stress, they will be better at controlling them.
- **Talk to friends and family**: They can supply emotional support and motivation to take action.
- **Identifying the trigger**: It is usually not possible to keep away from the trigger of stress. However, taking notice of unique triggers can help an individual strengthen coping and administration strategies, which may also include reducing risk.
- **Exercising regularly**: The physical undertaking will increase the body's formation of endorphins, which are chemicals that reduce stress. Exercise may include walking, cycling, running, exercising or participating in sports.

- **Mindfulness is trying**: People who practice this form of meditation use respiratory and thought methods to create awareness about their bodies and surroundings. Research suggests that mindfulness may have an effect on stress, anxiety, and depression.
- **Improvement in quality of sleep**: Too little sleep or negatively satisfying sleep can contribute to stress. Try to get at least 7 hours every night, and set common examples for sleeping and waking up. Avoid caffeine, eating, and intense physical entertainment in the hours before bedtime.

Treatment

If such strategies described above are no longer helping, it is important to see a healthcare professional for recommendation and support. A therapist may additionally advocate psychological therapy, such as cognitive-behavioral therapy (CBT).

A stated goal of CBT is to help humans cope with constant stress. In structured sessions, the therapist works to allow a person to change their behavior, thoughts, and feelings in relation to stress.

CBT may additionally aid in enhancing equipment and coping mechanisms to manage stress responses.

Occasionally, physicians recommend medications to deal with some symptoms of persistent stress. For example, they may prescribe antidepressants to combat anxiety or depression. For people with sleep disturbances, physicians may also prescribe sedatives.

When to see a doctor

Do not try to deal with chronic stress alone. If the self-help strategy is not working, a physician can present help and recommendations regarding medical options. They may also refer a man or woman to a specialized healthcare provider, such as a psychologist or psychiatrist.

Anyone overwhelmed with stress should see a health practitioner as soon as possible, mainly if they are using pills or alcohol to have suicidal thoughts or coping.

Take away

Stress is a common segment of everyday life. Short-term stress is usually harmless, but when it lasts and becomes chronic, it can cause many symptoms. It can contribute to the improvement of physical and mental disorders.

Self-help techniques include detecting triggers, increasing coping and avoidance strategy, achieving pulse and family, and practicing mindfulness.

If these techniques do not work or if stress is becoming overwhelming, a character should talk to a healthcare professional.

Chapter 8.

Core Mindfulness.

Fact and general experience are the outlines of the practical mind. When you think logically and rationally you are in realistic thought. This intellectual or scientific state of thought defines truth in facts, numbers, equations or phrases of cause and effect. Whether balancing your checkbook, baking cakes or doing crossword puzzles, you need the expertise of a discerning mind.

Proper consideration is essential for skilled skills. You need to understand how a skill works and when it's time to use it. A proper mind is a repository of information, which helps you to purposefully describe a problem and decide on a solution. To use the skill, you want to identify what one of the types of competencies are, and how to call them forward when you want them. For example, to regulate emotions efficiently, you need to be able to identify them, understand what activities and interpretations definitely indicate emotion, understand what feelings are like, what thoughts you need to do, forces, and their consequences. The more the extra exercise makes sense to you, the more and more you become enhanced in proper thoughts.

Proper thought is a lot less complicated when you are healthy, strong, calm, relaxed and fed, but much harder when you are sick, weak, stoned, tired or hungry. When you are feeling burdened or not feeling well, you begin to notice feelings.

However, sensible thinking is necessary to deal with reality, an emotional aspect in many issues of life.

Emotional mind
If rational thinking runs "cool" then emotion thinking runs "hot". Emotional and extreme reactions in the emotional mind make proper, logical thinking difficult. When an emotional kingdom controls your thinking and behavior, the idea of emotion is gone. The way you are understands how emotion behaves. Emotion thoughts can fill your system with angry strength or chant your power in depression. The way you are acting wisely can manipulate the development of clutter, feelings of loss. and more and more problems. Emotion ideas become careless, goofy, impulsive and impatient. Strong ideas distort facts, increase excuses, and reduce their understanding of the consequences.

Of course, some amount of emotion thought can be beneficial. Intense love is an inspiration for intimate relationships. It makes us want to inspire intense devotion or to live with very challenging tasks and sacrifice oneself for others. Mothers who are jogging through the fire to save their young are spirit minded. High in sentiment thought are often passionate about people, causes and beliefs - these are the dramatic, pleasant human beings of the world.

Grief and sorrow are emotional reactions to the troubles encountered in housing that make a difficult scenario worse. However, you can learn to use emotions and logic together to enhance your excellent life. This is the intention of the discerning mind.

Smart mind

The discerning mind is a lively integration of the emotional mind and the proper mind. The sensible mind collectively brings sensitivity to realistic thinking and emotional thought into a relaxed state. When you relate your troubles (what hurts you) to your troubles (what hurts), you are in a skillful and sensible mind. A sensible mind joins what forces you to your problems. Your effort to connect grief and sorrow to your issues with your proper, logical abilities is the foundation of skill and intelligent mind. But a sensible mind is more than this; the magic of the discerning mind is intuition.

Intuition is aware of the meaning, significance or truth of an event, in addition to intellectually analyzing it. Such intuitive emotions combine emotional experience and logical analysis but transcend them. Sometimes emotions can flow as intuition. You feel positive "know." If it is comfortable to "know", then it will still be valid when examined except only in the sense of the moment. A quite simple act validates your intuition, letting you know if you are sure whether it is emotionally biased or really intuitive. Intuition has symptoms of direct riding and immediate recognition with roots of reason and experience.

As you use your skills, you analyze how to act wisely. The sensible mind is comparable to intuition (or, perhaps, instinct is comparable to the sensible mind). By acting effortlessly with a discerning mind, you believe that you feel more than what you know. It is flexible, imagined, and open-minded. Awareness, in general, nourishes the discerning mind. Polarized thinking and an inflexible worldview interfere with the discerning mind.

One way to become intelligent is to be intelligent. Dedication is required to master such exercise. Practice identifying and fixing issues in your life. Some of the problems are, of course, quite easy and include writing a to-do list on your plan and a few extras besides the following. Everyone has problems in life. Successful humans are better at accepting that they cannot and change the matters that they can. These successful humans are informed with knowledge, equipped with experience, and guided through intuition. Be willing to take life's troubles together with your experience, knowledge, and intuition.

Wisdom grows when you use your senses, broaden your mind, and check skills. Wisdom works. Begin by studying core mindfulness skills, interpersonal effect skills, emotion law skills, and clear tolerance skills.

Intelligence, sensible mind or smart understanding depends on the integration of all ways of understanding something: knowing with the help of observation, understanding logically, knowing what you ride in your body (kinetic and sensory experiences), experience by learning and learning, and understanding through intuition. Knowing in these ways develops with cognition as you observe, analyze experience, learn, and intuit. By exploring multiple perspectives of understanding and turning to additional information, you enhance the experience of completeness, continuity, and coherence.

Wise Mind is like using a bike, which takes effort, balance, and steering. You can analyze the discerning mind, like you learned to ride a bike, only through experience. Just like you need to paddle to get started on a bike that needs a little more effort to make up your mind. As it would fall on you if you were going way too long for one aspect or another on the bike, find sensible minds by the way you try to find the consistency of your emotions and thinking. Intuition leads you to a sensible mind. Horrible emotions (anxiety, depression, anger, shame, and guilt) are similar to being put on a break. As you can climb and ride a bike, you can study and perform brilliantly to understand the clever idea.

As you endure the difficulties of living in the shortcomings of Wise Mind, you will achieve mastery. The feeling of being mastery, equipped and under control, does not mean that you will not make mistakes. Mastery is the approach to bring your skills to your problems in life.

Sometimes you can access enlightenment when faced with another character and remain calm under pressure. Sometimes you additionally find that in the midst of a disaster you instinctively understand the proper things to do. Sometimes approaching difficult trouble creates an insight that opens an inner door. Sometimes a sensible mind is looking at the whole picture, not just the parts. Sometimes when confronted with a challenging dilemma, Wise Mind demonstrates a clear alternative.

You will strengthen self-agency and self-awareness as you increase mindfulness, change your feelings, be tremendously reciprocal, and tolerate distress. Self-agency is the feeling when you have total control of your behavior. Instead of just realizing their behavior, self-agency is the master of conduct and takes responsibility for it. Self-awareness is the experience that you have when your various roles, emotions, attitudes and intellectual states come together and heal.

Meditation and intelligent mind

Mindfulness workout routines such as meditation promote internal peace, emotional control, perseverance, and a strong sense of self. Harvard Medical School MD Herbert Benson was one of the first to see the therapeutic value of meditation. He discovered that meditation can be called a "relaxed response", which is the physiological opposite of stress and anxiety. Meditation corrects stress loss. One of the easiest meditation workout routines is to observe your breathing. When you pay attention to your breath, you can find a sensible mind in the body core set behind your breathing. You can improve your ability to find your middle during meditation and become well aware of this quiet area. If you learn where your middle is and how it feels, you can go to this place, assured that you are answering in a sensible mind. Although meditation may also be unfamiliar, you can cultivate the ability to contemplate.

Meditation and Mindfulness strengthen your ability to focus on what is going on within yourself in any given situation. One way humans ride normally is to retreat from one's thoughts and feelings and to release your attachment to these mental events. Here the motives are based on your thinking and feeling, independent of the circumstances, seeing what is going on in one's mind like staring at the clouds in the sky.

Smart mind

Wise mind is calm. It is usually quiet and peaceful. When exchange or acceptance is an important sensible idea, there is peace in knowing which route to take. The sensible mind is now trapped in all-or-nothing thinking and can focus on what is good and functional.

In the discerning mind, you are in control of the emotional mind. Behavior is no longer established (controlled through one's emotions) though at the bearer of internal knowledge. Ultimately, this barely indifferent approach achieves the goals of self-observation and self-description. In fact, you can experience intense emotions like anger or anxiety while remaining in a sensible mind. It takes a lot of exercises - for everyone. Likewise, in the discerning mind, you are able to gain entry into the proper mind's expertise.

A sensible mind is courageous, that is to fear, but does whatever is necessary for the situation anyway. Desire is doing what is necessary for every situation. Desire requires courage again and again.

A sensible mind is full of confidence: Self-confidence makes you understand that there are certain issues in your life that you can cope with. When the sensible mind becomes clear, fear disappears. The sensible mind knows that you are doing fantastic and you can come down from the circumstances. When you strategize on your problems because of your troubles and crises, you are in a sensible mind and you do what you can satisfactorily.

Clear and consistent experiences of self: In a discerning mind, you can maintain your very own feelings, thoughts, and selections when others are around.

Self-description, a core mindfulness skill, helps you identify yourself and make unhealthy urges consistent with social stress and trade one's mind. At the same time, when you warrant your idea, you are not strong with ideas of choice. When you are calm you can additionally see who you are.

Self-observation develops self-understanding and accurate perception. Doing something contrary to or thinking "automatic pilot". The sensible mind is "being done in existing on purpose."

Through Mindfulness, you will boost your discerning mind's abilities to gain entry. When you solve the trouble with Wise Mind, you will enhance the splendor of your life.

Chapter 9.

Distress Tolerance.

You are at your emotional breaking point. Perhaps the worst has happened, or it may have been just the "last straw". The DBT sore tolerance capability you want is TIPP. This skill is designed to take you below the leadership of the metaphor (hopefully no longer literally).

TIPP means temperature, intense exercise, breathing, and muscle relaxation.

Temperature
When we are disturbed, our body regularly feels warm. To counter this, splash your face with bloodless water, grab an ice cube, or blow up the car's AC on your face. Changing your body temperature will give you both physical and emotional coolness.

Intense exercise
Exercise excessively to fit your excess emotion. You are no longer a marathon runner? That's right, you don't want to be. Sprint down the street, leap into the pool for a few laps or do jumping jacks until you exhaust yourself. Increasing oxygen glide helps reduce stress levels. Also, it is hard to stay dangerously upset when you are tired.

Emptied out
Even something as simple as keeping your breath under control can have a profound effect on reducing emotional pain. There are several types of respiratory exercises. If you have a favorite, breathe it. If you don't, try the

approach referred to as "box breathing". Each interval of breath will last four seconds. Take air in for four seconds, conserve it in four seconds, breathe four and conserve four. And start again. Continue to the center point on this respiratory specimen until you feel extra calm. Stagnant respiration reduces your body's fight or flight response.

Threaded music connection
The science behind paired muscle relaxation is truly fascinating. When you tighten a voluntary muscle, relax it, and enable it to relax, the muscle will grow more relaxed than it was once tightened. Relaxed muscle tissue requires less oxygen, so your respiratory and coronary heart rate will slow down.

Try this technique. Focus your attention on a group of muscles, such as the weight of the muscles in your arms. Tighten the muscles a great deal for five seconds. Then let the tension go. Allow the muscles to relax, and you will also begin to relax.

- **TIPP**
- **Temperature**
- **Intense exercise**
- **Breathlessness**
- **Relaxes muscles**

Crisis handling skills in TIPP will bring you closer to smart minds, where you will be able to make positive choices and face the product.

DBT Agony Tolerance Brief Introduction ACCEPTS is a team of competencies to help you tolerate a bad feeling until you are able to address and resolve the situation sooner or later. In the early season of the 90s sitcom *Friends*, Monica chose Pete Baker. He calls her out of town and says, "We need to talk." Does Monica wonder if this is two proper conversations or a terrible thing? He is in a psychological crisis ready for his return. The skillset she uses while waiting for Pete to come home is ACCEPTS.

This DBT ability stands for Activities, Contribution, Comparison, Emotions, Push Away, Thought, and Sensation. These strategies are designed to keep your thoughts manageable until you can get to the bottom of the problem.

The activities
Engage in an activity, and it can be about any healthy activity. Make strawberry jam, read a book, go for a walk, take your friend's name, wash dishes. Anything that keeps you busy and keeps your mind away from negative emotions will help. And when you finish, move on to a new activity. (You want a potentially very productive day, looking ahead of that dangerous situation!)

Contribution
Type something for every other person. Being a provider can relieve you of emotional distress in some ways. A function of the provider is also an activity which, as stated above, will help to clear your thinking about the problem at hand. Additionally, we feel accurate about ourselves when we help someone else, and this in itself can help you deal with stress. Help cook dinner, mow the lawn of a neighbor's or bake cakes for a friend or relative. Each of these ideas will distract your mind from your current situation.

Comparison

Keep your lifestyle in perspective. Is there a time when you have faced extra difficult challenges than today? Probably not - perhaps this is the most extreme situation and the most serious feeling you have ever experienced. (If so, you may additionally want to jump on the TIPP section again.) If this is the case, is there any other person who suffers more than you? Are you in your safehouse while in another part of the world someone is looking for food and shelter after another natural disaster? The purpose of this practice is no longer to add more grief and emotional pain to your current situation. Instead, use this ability to add an extraordinary approach to what you are experiencing right now.

Feelings

You have the energy to invoke a spirit contrary to your contemporary distressed feeling. Practice meditation for 15 minutes if you feel anxious. If you are feeling depressed, go to "Adorable Puppies" beforehand and Google image search. (If you need a real laugh, search for "ugly puppies".) Adding a dose of opposite emotion helps limit the intensity of the bad feeling.

Push away

When you can't deal with something yet, it's okay to temporarily take the problem out of your mind. You can distance yourself through distracting with other activities, thoughts, or arbitrariness. You can also set a time to return to the issue. You understand that this will be addressed, and you can stay away in the interim.

Ideas

Replace negative, worrying thoughts with activities that engage your mind, such as saying the alphabet backward or doing a sudoku puzzle. These deflections can help keep you away from self-destructive behavior as long as you are in a position to achieve emotion regulation.

Sensation

Use your five senses to calm everyone in their time of crisis. A self-soothing treat should include a lavender bath bomb and a hot bath with soothing music, a relaxing snack or watching your favorite show. Anything that ties to your senses can help you deal with the current situation.

Whether the situation is small (you just broke your shoe) or huge (you just broke your leg), many times it will happen that you have no control over an unpleasant event. During these times, you seek grief tolerance to make it through the forbidden landscape, engaging in unhealthy behavior. Intense thoughts do not stop forever. You can use the dialectical behavior ability to tolerate thoughts until the intensity decreases.

Reform means imagination, meaning, prayer, rest, a thing in the moment, leisure, and encouragement.

Imagination

Think about dealing with this problem successfully, being smart-minded, and realizing the feeling of achievement when the scenario is over. By doing this, you may actually be able to alter the result of the trouble in your favor.

Meaning

Try to find meaning in painful situations. What can you gain from this experience? Maybe you will be more empowered.

Maybe you will build new relationships. Maybe it will take you on a journey. Find a reason or a viable reason to state your current misery.

Pray
Prayer can come in any structure that works for you. Prayer can be fto God or any great power with the universe. Surrender your issues and ask them to tolerate the state of affairs for a while.

Rest
We become hurt due to painful conditions affecting our combat or flight instinct. Engage in pleasant things to calm the psychological crisis you are facing. These tasks may include deep breathing, yoga, hot baths, and enjoyable walks.

Be in the moment through past and future goings. Adding historical troubles to the situation, or tripping about the practical consequences of the situation in future, will not be useful in solving the problem. To do this, find a problem and focus all your attention on that task. One-track thought helps to experience ideas very rarely.

Holiday
On the right holiday, you can take the loss from all your stresses and return home equipped to take on the challenges left behind. Unfortunately, most of us are not in a position to take an actual vacation in a crisis situation. Instead, you can leave for a vacation in your mind. Imagine yourself elsewhere, such as a night stroll around the lake or driving on the Pacific Coast Highway. Stay longer on your "holiday" as needed, and rotate regularly as needed.

Hopefully, you will better "return" to the situation you are tolerating.

Encouragement
Incentives do not have to come from any external source to be effective. Encourage yourself through repetitive phrases such as "I got it", "I can amplify the moment". Say it out loud, say it proudly! You will be amazed at your ability to motivate yourself to make it through a challenging time.

- Improve
- Imagination
- Meaning
- Pray
- Rest
- An issue at the moment
- Holiday
- Encouragement

Grief tolerance strategies can be used anywhere, and each time you need to endure situations that you cannot change. Practice these techniques in mild situations and they will come naturally to you when big issues arise.

It can be difficult to make a sensible selection, especially when you are not of a sensible mind. Dialectical behavior therapy suggests the use of a pro and con list to weigh the consequences of your decision.

It is common for self-harm conduct to occur in emotional distress or to interact in various self-destructive conduct. If you want to act on an insistence or make a pro and con list to decide, it can be as simple as a few bullet factors in your mind or you can dig deep and make a long list on paper.

At this moment, which conduct is high-quality for you? This ability can be beneficial in hostile impulses and their terrible consequences.

Another simple way to increase your crisis tolerance in a crisis situation is to use your body's senses. Self-soothing through the senses can reduce the intensity of fastidious feelings.

Vision
Use your eyesight to focus on something else. Calculate how many places you can see a certain color in a room or the focal point on the texture of an object. You can also drag your phone and scroll through some of your favorite photos.

The hearing
Listen to the voice - any voice. Can you hear birds chirping or traffic noise outside? Turn up the volume on your favorite track and just listen. If you select soothing sounds, there are many apps that you can set on your smartphone on the go.

The taste
A short treatment can compel you to pay attention to something that you are going through at a challenging moment. You want a complete meal - a piece of gum or some mint together.

Touch
How to feel a pen in your hand, run through your hair with your fingers, or embrace your sense of touch through the

use of a fidget toy. When appropriate, you can take a hot bath or wrap yourself in a blanket.

The smell
Whether it is good or bad, the center of attention on the smell of something is in the air. Can you feel the smell, or know its components? To reach an odor that you seek to soothe, put a few drops of your favorite essential oil on a cotton ball and put it in a plastic bag with you.

Protest
While you are technically only five senses, the DBT introduces the sixth sense of movement. Your emotional state can be changed through your body movements, so take a walk around the block or dance to your favorite song!

- Reduce self-agony
- vision
- Hearing
- The taste
- Touch
- The smell
- protest

Your sensations are a device that you usually have to reduce the intensity of a situation. To incorporate mindfulness into self-soothing skills, try to focus on only one experience at a time.

Sometimes you will have an undesirable situation that will not change. Now you may not like it or accept it, although acceptance will enable you to provide peace and supplies to move the region forward.

Fundamental acceptance recognizes that we all have choices, and it comes down to us from time to time to decide whether we are being given the reality of our situation. You can choose to be apathetic about the situation, or you can choose to distribute it and move on.

Imagine being terrified of the dentist. You tried to leave it. You tried to deny it. But you recognize that you have a cavity. You had a proper relationship with your previous dentist, but he just retired. Your new dentist is not hot with his victims and is very eager to get around that fast, spinning drill.

In an effort to avoid the dentist and manage pain, you start cutting out some of your favorite foods, which cause irritation in the cavity, such as sugar and cold food. But it's okay, isn't it? You are thinking that you must eat more fish and vegetables. This is fine when you consume an unknown cavity irritant, or the cavity pain flares up for no real reason.

By training radical acceptance, you choose to accept that you are afraid of the dentist, it will be a frustrating experience, and the cavity has to be cleaned regardless. You can't skip it or you'll eventually need a root canal, and no one has time for that. (Read: The Most Great Dental Methods As He Is.)

So you go to the appointment, prepared for the worst, and forty-five minutes later you walk out with a full set of enamel and a renewed commitment to flossing.

And now don't forget this - when you're in that dentist chair with a suction tube from your lips and the hydrologist is washing your entire face with water, you use the prescribed impact capability to make it different sides correctly.

DBT's Trouble Tolerance Talent training may also appear daunting, but emotion legislation gives you more control over the urge to engage in impulsive behavior. Whether you have a mental illness or disorder or not, you can increase your mental fitness and ability to tolerate critical situations. A DBT application or DBT practitioner can help you develop these skills.

About Occupational Therapy (DBT)

Dialectical Conduct Measures (DBT) is talent training to deal with a myriad of difficult situations. The dialectical behavior measure was initially developed as a cognitive behavioral therapy replacement in the treatment of borderline character disorder (BPD). However, it is now considered the popular gold of medicine for many mental fitness issues. DBT has been successful in treating substance abuse, traumatic stress disorder, binge urge to consume or cleanse, and others.

Chapter 10.

Emotion Regulation.

Many humans with Borderline Personality Disorder (BPD) battle with simple emotion rule skills. In fact, one of the foremost researchers in BPD (Marsha Linehan) and developer of dialectical behavior therapy (DBT) for BPD, has suggested that emotion regulation is lacking at the core of the disorder. We know that modifying our thoughts is important for everyone, whether or not we have BPD, but what is emotion regulation, and how can you hone your emotion rules skills?

Understanding emotion regulation

We all consider feelings to be either bad or good, depending on the day. As children, most of us typically research how to manage, express, and cope with these emotions. For some humans, however, with those with BPD, emotion regulation is very difficult, from time to time due to traumatic childhood experiences, abuse, or trauma, and from time to time because it has not been proven so far.

Although this is considered to be an essential area of detection in the world of psychology, there is no consensus on the definition of "emotion regulation" over a period of time. Many researchers define sentiment legislation as the ability to increase or decrease their views as necessary.

For example, if you feel nervous in the middle of a meeting at work, you can try to calm yourself by inquiring about something else.

Other researchers use a more comprehensive definition of emotion regulation, viewing it as a set of abilities that help your emotional system to be healthy and functioning. Since emotions are no longer complete and permanent, we can do research to control what emotions we have, how much more it is, when we have it, and how we react to it. The defining factor of emotion rules is that it occurs when an objective is activated. The goals are relatively personal. They are what we put into our heads - the way we like things. Your desires can be activated through your environment in a whimsical or subconscious way, including people, objects, images, words, and sounds.

Examples of emotion regulation
Here are some examples of active dreams that trigger the formation of your emotions:

Influencing a choice in someone else: If you are a parent, your intention may also be to aid your baby's analysis of how to modify his or her feelings. When your toddler has a meltdown, you may feel angry or even happy, but alternatively, after shouting or laughing, you modify your feelings to discuss calmly with your child how he should react instead. This is known as external emotion regulation.

Influencing an exchange in yourself: If one of your goals is to be more positive, you can revise your terrible thoughts by focusing on brilliant people. Regulating one's own thoughts is called internal emotion regulation. Sometimes this form is pushed with the aid of the law as to what our tradition sees as accurate or terrible feelings, or how we should behave in certain circumstances, such as funerals.

Meeting Long-Term Goals: You can also control your emotions to achieve any other goal. For example, when your boss treats you poorly at work, you act as if it won't bother you because you're expecting a job promotion.

Changing duration, intensity or type of emotions: Many times, we work to increase or decrease the depth of our emotions. For example, you may also feel depressed or anxious, but no one at work knows about it. We additionally alternate how long our feelings are. An example of this is that you are no longer thinking about the anxiety you feel over financial difficulties and instead want to keep your thinking busy with other activities. At different times, we can also alternate the types of emotions we feel. If you fell in front of everyone, you can choose to make a joke about it instead of being embarrassed.

Subliminal Regulation: This type of emotion regulation also prevents you from knowing or feeling it. An example of this would be to rapidly switch channels when something disturbing is being shown on the TV.

Sometimes these targets overlap. For example, you can happily talk to a baby (external) to help ease your personal anger and frustration (internal).

Emotion regulation process model
The prevailing theory of emotion rules is known as the system model. First, our emotions arise through a

"situation, attention, evaluation, reaction" sequence, such as:

Status: Emotion technology begins in a situation. It may additionally be an external condition, such as a friend making an integral comment, or it may be a perception or emotion in your own head.

Note: This scenario attracts your attention. For example, your attention may be drawn to the way your pal crosses his palms as if he is angry.

Evaluation: You evaluate the situation. In this instance, you may also be concerned that this person no longer wishes to be your friend.

Feedback: Your initial reaction may be physical and/or emotional. Your face may additionally flip crimson and you may also experience an injury. Then you reply to the person, who can also trade the state of affairs and start a new sequence again. For example, you inform your friend that the comment has hurt your feelings and ask why he said it. He can then express regret or say that he is having a terrible day.

When we talk about emotion regulation, we can choose any part of the above sequence of emotion constructions and influence our emotions accordingly.

Situation: We can choose to keep away from humans or situations that we believe will detect injury, interact in situations we find positive, or we can change our behavior by alternating the status of those cases.

Note: We can focus on something else in the situation, such as the nonverbal sign giving different man or woman or maybe backward to what he is saying.

Valuation: We can trade the way we are thinking about the situation. For example, if the emotion generation sequence was started by you saying "I'm so stupid," you can tell yourself that this is no longer appropriate and it's just a feeling that you're doing right now. In the example above, after being stressed that your friend no longer wants to be your friend, you can remind yourself that you are going to the conclusion and a necessary comment does not mean leaving your friendship.

Feedback: Alternatively how can we react to the situation? Instead of being inactive and exiting, you can do some respiratory exercises. Rather than overcome an uncomfortable situation, you can trust your pal with you. Instead of overreacting to what someone says, you can ask him or her more and more so that you understand what they mean.

Factors in healthy sentiment legislation include:
The ability to recognize that you are having an emotional reaction and to understand what the response is.

Accepting your emotional reactions as an alternative to dismissing them or reacting with fear. This can also be difficult for humans who do not have BPD, as feelings such as anger or sadness are regularly discouraged with the help of society.

The ability to access technology that allows you to reduce the intensity of emotion that you want. If this ability has offended you, you are not supplied with the desire to physically whip at them or hurt a long line of their way. In fact, when you are upset, you should be in a position to negotiate in goal-directed conduct if you are well versed in emotion regulation.

The ability to manipulate impulsive behavior when you are upset. If you feel that you have received disturbing news due to your domestic disturbance, you can curb the tilt by throwing it completely to the floor or punching a hole in the wall.

Because people with BPD can battle with some or all of the skills in this list, this broad definition of the spirit legislation outlined above is most useful in describing the lack of rules present in BPD. Fortunately, it is feasible to increase some emotional law competencies that you are lacking.

How to improve the ability to regulate your emotion
If you need to work on your emotion legislation skills, this education is enjoyable with the help of a BPD therapist, noting that it is an essential component of the dialectical conduct measure for BPD. However, you can try some workout routines of your own therapy such as:

Reducing Emotional Vulnerability: Get plenty of sleep, eat a nutritious diet, stay active, and take time to do activities you love that keep you away from some emotional ups and downs associated with BPD.

Mindfulness: Mindfulness is currently an exercise and is one of the core capabilities in DBT. Wise people analyze being conscious of the breath they take, the tension in their muscles, and even their pulse rate. They cut their food slowly and deliberately, and listen to their bodies for cues that they are full. They can see themselves in difficult moments, confident that these examples will also pass. Being alert can help you learn how to use healthy coping abilities to deal with your emotions.

Emotional Acceptance: Emotions in and of themselves are no longer desirable or bad, although they can be scary, especially when they are intense. Learning to accept your thoughts takes practice, although the more you do it, the more normal it becomes. Mindfulness capabilities can also help with this.

Chapter 11.

Interpersonal Effectiveness.

Interpersonal effectiveness, at its most basic, refers to the ability to engage with others. This includes the qualifications we use:

- Participate in relationships
- Remaining preferences vs. demands
- Balance "wants" and "shoulds"
- Build a sense of mastery and self-respect (Vivaan, 2015)

Our ability to interact with others can be damaged through the intention we are thinking about our interactions. There are three primary dreams to negotiate:

- Achieving our purpose
- Maintaining our relationships
- Keeping one's self

Each intention requires interpersonal skills; while some interpersonal skills will be applied in many situations, certain skills will be particularly necessary to accomplish one of these goals.

When we are working closer to achieving our objective, we want capabilities that include clarity on who we favor with the conversation, and identify what we want to do to achieve the results we want.

When maintaining our relationships is our first priority, we want to understand how accurate relationships are to us, how we desire the person to feel about us, and what we need to do to maintain the relationship.

Finally, when we intend to hold on to our self-esteem, we will use interpersonal abilities to help us feel the way we want to feel after the conversation and from our values and reality. (Vivian, 2015).

Interpersonal influence and dialectical behavior therapy
The main focus of dialectical behavior therapy (DBT) is interpersonal effectiveness. In fact, it is a 2D core capabilities module in traditional DBT, with lots of content and assets to improve the client's interpersonal skills.

You might be wondering why the reciprocal effect is so important that it warrants an entire module in one of the most well-known forms of medicine. Certainly, verbal exchange is important, but does it, without a doubt, require a lot of time and effort? Why?

DBT believes that these capabilities are very essential due to the fact that the way we talk with others has a tremendous impact on the extraordinary and the impact of our relationships with others (Bray, 2013). In turn, the brilliance of our relationships and the impact of our interactions have an impact on our well-being, our understanding and confidence, and a lot of perception of who we are.

While there are many capabilities related to interaction and interaction with others, DBT focuses on two major components:

- The ability to ask for the things you want or need
- Ability to say no to requests when appropriate

The importance of developing your interpersonal influence skills

By now, you have just diagnosed the importance of being good, or at least good enough, mastering conversation and skill. However, you may be think that if you have the ability to talk with others with minimal effectiveness, then you are set! Why bother working on the skills you already have?

Like any complex skill set, by no means will there be a point at which you have fully mastered them. Even the best motivational audio systems and family members of public experts are no longer ideal communicators. There is usually room for improvement!

Research has supplied evidence that increases these interpersonal competencies, particularly for clients with borderline personality disorder (BPD) leading to fine results. For example, DBT talent use has been shown to increase BPD symptoms, reduce affective instability, and improve a client's relationship abilities (Stepp, Eppler, Jehung, &Trull, 2008).

Games and activities (for groups) to develop effective interpersonal skills

However, there are many worksheets and male or female workout routines that you can engage to build your interpersonal skills, they are not always the most overwhelming way to do this. It is no surprise that interacting with others is a way to increase quality!

Not solely team activities, which are generally better at improving interpersonal skills, they are often extra fun. Below, we have listed and described five fun video games and activities that you can use to enhance your interpersonal influence (as a handout you can use to check your interpersonal skills).

Skill assessment handout

Before attempting to enhance your interpersonal verbal exchange skills, it is a top concept to find out where you are currently with each one. Evaluation on page three of this handout can help.

Here, you will find 29 skills, such as:

- Talk about yourself
- Hear what the person says
- Listening - showing interest in people
- Reply to praise
- React to negative feedback
- Self-disclosure as appropriate

For each skill, you look forward to rating yourself on a scale of 1 to 5, according to the following rubric:

1 - I am terrible at that skill
2 - I'm poor
3 - I'm good sometimes
4 - I'm usually good
5 - I'm always good

You can take the common of your scores to provide a simple "interpersonal effectiveness" ability rating of your own, although male or female scores are precious through themselves.

If you are looking to improve your conversation skills, be positive enough to establish a baseline first. If you have a baseline to compare and return to, it is very easy to note the corrections!

Do not try to listen to the activity

In this fun and carefree eye-opening activity, team contributors will be at risk to put their looking chops to the test.

This activity requires the team to break up into pairs. In every pair, a man or woman must be unique to speak "hears" to play another role for the first time.

It is advisable to speak directly to the first speaker (partner A) for two minutes, whatever concerns they wish to discuss. While partner A is talking, partner B's job is to make it clear that he is not listening to partner A.

Partner B cannot say anything; they can only use an alternative dependent on paper language to talk to Partner A on his message.

At the end of the two-minute conversation with partner A, partner B receives two minutes to talk while partner A "listens"

The team will potentially be able to know that when their teammate is not listening, it is exceptionally difficult to keep up with the speaking! This is an important lesson of the activity: that body language plays an important position in communication, and listeners have a considerable influence on how interactions other than speakers operate.

Once all crew members take both the turn of Speaking and "listening", each person must write their immediate responses to be the speaking partner that is no longer really heard.

They will probably come up with feelings such as:

- **I was disappointed.**
- **I used to get angry.**
- **I felt that I was not important.**
- **I felt like what I was saying was boring.**
- **I could not talk.**
- **I felt insignificant.**

Next, crew members should observe behaviors that their partner was exhibiting to show that they are not listening, such as:

On the far side, tilting the head towards the floor or turning to the side, avoid:
- Looking at the floor

- Bent arms / crossed legs
- Blank or bored expression
- Yawning, whistling, scratching, or isolated activity with active listening
- Forethought (when searching in one's environment or one's phone, etc.)
- No chat

While this exercise is certainly an exaggeration as to what it is to talk to someone who is not listening, it can help those who are now able to display their personal conduct while interacting with others. They are not very attentive or limited in their social abilities.

It is easy to decide to practice live listening in your interactions, but it is harder to maintain all in keeping with the target behavior (and of course the non-target behavior). Practicing this will help contributors to choose the behaviors that make a person a good listener.

You can find this workout on page 4 of the handout mentioned above (interpersonal skill exercises).

Sabotage exercises

This is some other enjoyable workout that adjusts negative interpersonal behaviors to highlight accurate interpersonal behaviors.

This exercise should be performed in a giant group, which is sufficient for damage in at least two or three agencies of 4 to 5 persons.

Instruct each team to spend 10 minutes brainstorming, discussing, and listing all the approaches they might consider to sabotage a group assignment. Whatever they may believe, it is an honest game - it simply aspires to be sufficiently disruptive for the proper handling of trains!

Once each team has a good list of ways to sabotage group assignments, once again accumulate to the larger crew and compare responses. Write them all on a chalkboard, whiteboard, or a flip board in the front of the room.

Next, teach improving groups and building 5- to 10-point contracts with agreed-upon suggestions for successful group work. Group participants should draw from subversion ideas (i.e., what the successful crew should do for work) to take appropriate ideas (i.e., what to do for successful group work).

For example, if a group "does not communicate with any of the different crew members" as a way to sabotage group assignments, they may come up with something like "different communicating with team members" often as a guiding principle for successful group work.

This exercise will help individuals analyze what makes for a positive team experience, as well as give them the opportunity for a profitable team journey along the way.

This workout was described on page 14 of this handout.

Strengths and weaknesses of the group
When it comes to getting people to work, groups have a very important benefit - they can overcome individuals'

weaknesses, supplement their strengths, and create balance in the group.

Individuals in the group will have some unavoidable thinking and conversation about their own strengths and weaknesses in this exercise, as well as the strengths and weaknesses of other group individuals and as a team overall.

To provide an effort to this workout, coach the group to make assumptions about the strengths and weaknesses of each individual team member. Encourage them to be honest with each other, especially when discussing weaknesses.

Once each group has come up with an accurate list of strengths and weaknesses for each team member, what each group believes will have an impact on the team's dynamics. Which forces will positively influence group interactions? Which weaknesses have the potential to throw the monkey wrench into team interaction?

Finally, every team talks about the structure of a "right" team. Does it have more to do with participants with similar characteristics or with wider differences of personality, abilities, and skills? What are the benefits and dangers of every type of team?

This discussion will help contributors to think critically about what makes an accurate team, how one of a variety of personalities interact, and their behavior, crew norms, or the different personalities and skills of others .

How to change expectations to match

This workout is also described on web page 14 of the handout on interpersonal skills (interpersonal skills exercises).

Calculate squares
This entertainment is an enjoyable and engaging way to inspire group entertainment and communication.

All you want is this photo (or comparable picture of several squares) to be displayed on a wall or board in front of a PowerPoint presentation or room.

In the first step, provide the group with a few minutes in my view, trust the variety of classes in the parents and write your answer. They should do this without speaking to others.

Subsequently, each crew member should report the diversity of the classes they counted. Write these on the board.

Now instruct each participant to find someone to pair with and memorize the squares again. They can discuss with each other when determining how many classes there are, although none are.

Once finished each pair should share their boundary once again.

Finally, participants have to structure four to five members every year and train them to count the number of extra time. Once they complete the task, each group is counted once more.

At least one team must have almost explicitly counted the correct diversity of the classes, which is forty. This group explained how the rest of the contributors crossed 40.

Finally, lead the entire crew through a dialogue of team rapport, and why the count (likely) kept getting closer and closer to forty as more and more humans collectively got around the problem.

Participants will examine the importance of accurate group communication, the practice of working in pairs and in groups, and with any luck will lead to a happy ending of this activity.

You can find out more and more facts about this effort.

The game is a happy twist on a historical classic - meeting a new character and introducing them to the group.

You design the game on the first day of a crew therapy, training, or separate recreation to take advantage of the possibility of presenting each group member.

The crew members are paired with a man or woman sitting next to them. Ask them to introduce themselves to everyone else and find something attractive or unusual about themselves.

Once each pair is connected and something attractive about the individual is located, return the focal point to the larger group.

Tell group members that each man or woman must match their partner to the group, although with a seizure - they cannot use phrases or recourse! Each associate needs to present the individual associate with tasks only.

The game is not only an excellent icebreaker to get people completely acquainted with each other, but it is also a fun way for crew contributors to see both usefulness for verbal interactions (something you can only identify when it cannot be used!) and the importance of nonverbal communications.

If you have the time, you can lead the team in discussions of nonverbal communication, the signals we choose in other people's behavior, and the way you are commenting on them is important.

You can do an additional investigation about this game here.

Ways to improve your mutual impact in the workplace
While there are many ways to work on your interpersonal skills, it is a little harder to discover ways to increase your work-specific interpersonal impact.

Luckily, most of these abilities transfer well from medicine to family life, interactions with friends, and the workplace. Additionally, there are some workout routines and sources that have been developed to improve interpersonal capabilities directly related to work.

Below you will find some different ways to improve your communication at work.

Interpersonal skills

This profitable handout can be reviewed over and over again as you or your mentor work on improving interpersonal influence.

It underscores the capabilities that are wanted to communicate effectively with others, divided into three of a kind talent sets:

- Objective effectiveness
- Relationship effectiveness
- Self-esteem efficacy

For each set, it is important to help you keep in mind what abilities are included.

For goal effectiveness, the acronym "DEAR MAN" is used:

D - Describe: Use clear and concrete words to describe what you want.

E-Express: Make others understand how a state of affairs makes you feel through expressing your feelings clearly; do not expect others to read your mind.

A - Actual: Do not beat round the bush - say what you want to say.

R - Reinforce: Reward humans who respond well, and beef up why your favorite outcome is positive.

M - Mindful: Don't forget about the goal of conversation; it can be easy to get caught up in dangerous arguments and lose focus.

A - Manifest: show confidence; Reflect on your posture, tone, eye contact, and body language.

N - Non-reserved: Anyone can have everything they desire out of an interplay at all times; be open to conversations.

These skills allow those who express their needs and desires to efficiently and concisely make them out of a conversation with whatever they want.

The acronym for relationship effectiveness is "GIVE":

G - Gentle: Do not attack, threaten or decide at some point in your interactions; get the occasional "no" for your requests.

I - Interest: Show interest in addition to interrupting the way the other character is heard.

V - Validity: Externally validating the other person's thoughts and feelings; accept their feelings, identify when your requests are being solicited, and recognize their opinions.

E-Easy: Have a convenient attitude; try to smile and act.

These competencies help people to maintain relationships with others.

Finally, the acronym for self-esteem effectiveness is "FAST":

F - Fair: Be fair; now, not only for others but also for ourselves.

A - Apology: Do not apologize until there is a need for it; do not apologize for forming an opinion, an opinion or disagreeing.

S - Stick to values: Do not compromise your values just to be liked or to get what you want; stand up for what you consider.

T - Truthful: Avoid dishonesty such as exaggeration, manipulation as appearing helpless or lying.

Did you know that can download the audiobook version of this book for free?

Click here for Audible US
Click here for Audible UK
Click here for Audible FR
Click here for Audible DE

Don't forget to leave a review on this book. It is simple! Just click on this LINK and you will be directed to the right page. It's very important for me. I'll appreciate if you do.

Made in the USA
Monee, IL
05 November 2021